Lecture Notes in Civil Engineering

Volume 104

Series Editors

Marco di Prisco, Politecnico di Milano, Milano, Italy

Sheng-Hong Chen, School of Water Resources and Hydropower Engineering, Wuhan University, Wuhan, China

Ioannis Vayas, Institute of Steel Structures, National Technical University of Athens, Athens, Greece

Sanjay Kumar Shukla, School of Engineering, Edith Cowan University, Joondalup, WA, Australia

Anuj Sharma, Iowa State University, Ames, IA, USA

Nagesh Kumar, Department of Civil Engineering, Indian Institute of Science Bangalore, Bengaluru, Karnataka, India

Chien Ming Wang, School of Civil Engineering, The University of Queensland, Brisbane, QLD, Australia

Lecture Notes in Civil Engineering (LNCE) publishes the latest developments in Civil Engineering - quickly, informally and in top quality. Though original research reported in proceedings and post-proceedings represents the core of LNCE, edited volumes of exceptionally high quality and interest may also be considered for publication. Volumes published in LNCE embrace all aspects and subfields of, as well as new challenges in, Civil Engineering. Topics in the series include:

- Construction and Structural Mechanics
- Building Materials
- Concrete, Steel and Timber Structures
- Geotechnical Engineering
- Earthquake Engineering
- Coastal Engineering
- Ocean and Offshore Engineering; Ships and Floating Structures
- Hydraulics, Hydrology and Water Resources Engineering
- Environmental Engineering and Sustainability
- Structural Health and Monitoring
- Surveying and Geographical Information Systems
- Indoor Environments
- Transportation and Traffic
- Risk Analysis
- Safety and Security

To submit a proposal or request further information, please contact the appropriate Springer Editor:

- Mr. Pierpaolo Riva at pierpaolo.riva@springer.com (Europe and Americas);
- Ms. Swati Meherishi at swati.meherishi@springer.com (Asia - except China, and Australia, New Zealand);
- Dr. Mengchu Huang at mengchu.huang@springer.com (China).

All books in the series now indexed by Scopus and EI Compendex database!

More information about this series at http://www.springer.com/series/15087

Job Thomas · B. R. Jayalekshmi ·
Praveen Nagarajan
Editors

Current Trends in Civil Engineering

Select Proceedings of ICRACE 2020

Editors
Job Thomas
Cochin University of Science
and Technology
Kochi, India

B. R. Jayalekshmi
National Institute of Technology Karnataka
Mangalore, India

Praveen Nagarajan
National Institute of Technology Calicut
Calicut, India

ISSN 2366-2557 ISSN 2366-2565 (electronic)
Lecture Notes in Civil Engineering
ISBN 978-981-15-8150-2 ISBN 978-981-15-8151-9 (eBook)
https://doi.org/10.1007/978-981-15-8151-9

© The Editor(s) (if applicable) and The Author(s), under exclusive license to Springer Nature Singapore Pte Ltd. 2021
This work is subject to copyright. All rights are solely and exclusively licensed by the Publisher, whether the whole or part of the material is concerned, specifically the rights of translation, reprinting, reuse of illustrations, recitation, broadcasting, reproduction on microfilms or in any other physical way, and transmission or information storage and retrieval, electronic adaptation, computer software, or by similar or dissimilar methodology now known or hereafter developed.
The use of general descriptive names, registered names, trademarks, service marks, etc. in this publication does not imply, even in the absence of a specific statement, that such names are exempt from the relevant protective laws and regulations and therefore free for general use.
The publisher, the authors and the editors are safe to assume that the advice and information in this book are believed to be true and accurate at the date of publication. Neither the publisher nor the authors or the editors give a warranty, expressed or implied, with respect to the material contained herein or for any errors or omissions that may have been made. The publisher remains neutral with regard to jurisdictional claims in published maps and institutional affiliations.

This Springer imprint is published by the registered company Springer Nature Singapore Pte Ltd.
The registered company address is: 152 Beach Road, #21-01/04 Gateway East, Singapore 189721, Singapore

Contents

Treatment of Well Water Contaminated with Septic Tank Effluent Using a Modified Compacted Sand Filter 1
M. Harikumar, P. Sikha, M. P. Amrutha, F. Jamshiya, and T. Arathi

Heavy Metal Fractionation in Aerobic and Anaerobic Sewage Sludge 11
Sooraj Garg, M. Mansoor Ahammed, and Irshad Shaikh

Environmental Remediation of Oil Contaminated Soil 21
A. Nishida, Aparna Gopinath, S. Chandraj, K. Radhika, and R. Sethu

Treatment and Reuse of Periyar Sedimented Soil Using Nanochemicals 35
B. Diya and Ann Mary Mathew

A Study on Red Soil to Form an Bouncy Cricket Pitch 45
S. Amritha and V. Rani

Influence of Flood on the Behavior of Friction Piles 55
R. S. Athira, S. H. Jasna, K. A. Renjini, Manjima Jayan, Shruthi Johnson, and J. Jayamohan

Feasibility Study of Using Coir Geotextiles in Permeable Pavement Construction for Stormwater Management 63
Mohan Kavitha, Subha Vishnudas, and K. U. Abdu Rahiman

Assessment of Effect of Filler in the Properties of Cement Grout 73
A. B. Kavya and S. R. Soorya

High-Strength Geopolymer Mortar Cured at Ambient Temperature . . . 83
Job Thomas and N. J. Sabu

Development of High Strength Lightweight Coconut Shell Aggregate Concrete 95
A. Sujatha and Deepa Balakrishnan

Comparison of the Performance Between Concrete Filled and Stiffened LDSS Column 105
Divya Roy and Milu Mary Jacob

Aspect Ratio Factor for Strength Correction of Pressed Earth Brick Prisms 115
Nassif Nazeer Thaickavil and Job Thomas

Numerical Investigation of Punching Shear Strengthening Techniques for Flat Slabs 123
Navya S. Ravi and Milu Mary Jacob

Investigation of Bolted Beam–Column Steel Connections with RBS Subjected to Cyclic Loading 133
Deepa P. Antoo and Asha Joseph

Effect of Shock Absorbers in Enhancing the Earthquake Resistance of a Multi-storeyed Framed Building 147
Deepa Balakrishnan, Anjali, and Salauddin

Review Paper on Pavement Condition Assessment 155
Saranya Ullas and C. S. Bindu

Land Base and Digital Elevation Model Creation Using Unmanned Aerial Vehicle 165
Anupoju Varaprasad, Kundangi Haritha, Shaik Syffudin Soz, and Samoju Chiranjeevi Achari

Multiphase Modelling of Orifice Cavitation for Optimum Entrance Roundness 185
V. R. Greeshma and R. Miji Cherian

Flood Risk Assessment Methods—A Review 197
Ginu S. Malakeel, K. U. Abdu Rahiman, and Subha Vishnudas

Flood Hazard Assessment and Flood Inundation Mapping—A Review 209
Reshma Antony, K. U. Abdu Rahiman, and Subha Vishnudas

About the Editors

Dr. Job Thomas Professor, Cochin University of Science and Technology is a renowned academician and structural consultant. He graduated from University of Kerala and completed his PhD at Indian Institute of Science. His area of interest are sustainability aspects in civil engineering, building materials, innovative construction practices, project management etc. He has 20 years of teaching experience and is the member of Indian Concrete Institute, Institution of Engineers and Indian Society for Technical Education. He received many national and institute awards. He received many research funds projects from AICTE, DST, KSCSTE etc. He has published more than 40 Scopus indexed papers in various international journals. He is the reviewer for many international journals.

Dr. B. R. Jayalekshmi is currently serving as Professor in the Department of Civil Engineering, National Institute of Technology Karnataka, Surathkal. She obtained her B.Tech. (Civil) from National Institute of Technology, Calicut (REC Calicut), Ph.D. from National Institute of Technology Karnataka, Surathkal and Post Doctoral Fellowship from Indian Institute of Technology, Madras. Her major areas of research interests include dynamic soil- structure interaction, seismic structural engineering and applications of finite element method in structural engineering. She has 25 SCI/SCOPUS publications and published more than 110 research articles in international journals and conferences. She is a member of Special Structures Sectional Committee of Bureau of Indian Standards and recipient of Women Achievers Award 2017 of Institution of Engineers (India), Karnataka. She has been a reviewer for Elsevier, Springer and Technopress journal articles and technical papers of international conferences.

Dr. Praveen Nagarajan had his Civil Engineering education from NIT Calicut and IIT Madras. After a brief stint as Bridge design Engineer at L&T Ramboll, Chennai, he took to academics. His areas of interest are reinforced and pre-stressed concrete, bridge engineering and structural reliability. He has published more than 90 technical papers in these areas and has authored the books 'Prestressed Concrete Design' (published by Pearson) and 'Matrix Methods of Structural Analysis' (published by CRC). He is the recipient of several awards like the Valli Anantharamakrishnan Merit Prize from IIT Madras, E P Nicolaides Prize from the Institution of Engineers (India), the Best Young Teacher award from NIT Calicut, ICI -UltraTech Award for Outstanding Young Concrete Engineer of Kerala by the Indian Concrete Institute (ICI) and ICI-Prof. V. Ramakrishnan Young Scientist Award by the ICI. He has guided 4 PhD students and more than 40 M-Tech projects. He is also guiding ten research scholars for their doctoral degrees. Presently, he is working as Associate Professor in the Department of Civil Engineering at National Institute of Technology, Calicut.

Treatment of Well Water Contaminated with Septic Tank Effluent Using a Modified Compacted Sand Filter

M. Harikumar, P. Sikha, M. P. Amrutha, F. Jamshiya, and T. Arathi

Abstract The flood that occurred in the month of August 2018 had brought severe damages all over Kerala. The major problem after the flooding was the contamination of well water with septic waste. This problem created a situation where proper drinking water was not available to the victims. A similar situation was faced by the residents of Velam Panchayath of Kozhikode district, where the well water was contaminated using septic tank effluent. Disinfection using bleaching powder was the only method adopted by the local authorities to make the well water potable. Since the septic waste contains toxic content and affects the human life significantly when consumed, an efficient and economic method of well water treatment is very necessary. The well water should be treated effectively after it is pumped into the overhead tank and then used for domestic purposes. This requires the designing of a filter in which the water gets purified in stages. The purified water coming out of the filter should be tested and made sure for drinking. The aim of this paper is to make a keen attention towards this problem and to implement some control measures to minimize this problem to certain extent by fabricating a Modified Drawer Compacted Sand Filter (MDCSF). This model consists of different drawers each filled with gravel, sand, activated charcoal, and silver-impregnated sand. Although the conventional drawer compacted sand filter has been used in the treatment of contaminated water, in this paper, a modification is made to the existing design by the introduction of forced aeration, using a silver-impregnated sand layer and an activated charcoal layer.

Keywords Septic tank effluent · Modified drawer compacted sand filter · Well water treatment · Coliform · BOD · COD · TDS

M. Harikumar (✉) · P. Sikha · M. P. Amrutha · F. Jamshiya · T. Arathi
Department of Civil Engineering, College of Engineering Vatakara, Kozhikode, India
e-mail: harikumar0907@gmail.com

1 Introduction

Contamination of drinking water sources by sewage can occur from raw sewage overflow, leaking sewer lines, and the application of sludge and partially treated wastewater to land. Septic tank effluent (STE) is the effluent discharged out of a septic tank. During natural disasters such as flooding, this effluent may get mixed up with well water, making them unfit for use and creating water scarcity problems [1]. The pollution and ill effects due to STE are not properly investigated or monitored. This study aims to fabricate a modified drawer compacted sand filter that relies upon different methods of filtration like straining, absorption, adsorption, biological action, etc. This process is undertaken by different layers of gravel, sand, silver-coated sand, and activated charcoal. Synthetic septic tank effluent is prepared under laboratory conditions and is filtrated through the Modified Drawer Compacted Sand Filter (MDCSF). The filtered water can be tested and compared with drinking water standards.

2 Objectives of the Study

The primary objectives of the study are outlined as follows:

- To synthetically produce the septic tank effluent (STE) in a laboratory.
- To check the chemical and biological characteristics of STE such as
  - Biological oxygen demand (BOD)
  - Chemical oxygen demand (COD)
  - Coliform content
  - Phosphate
  - TDS.
- To find the treatment efficiency at varying conditions of hydraulic loading rate and pH of drawer compacted sand filter (DSCF) with silver-impregnated sand.
- To get potable water from the filter.

3 Preparation of STE

Since it is inconvenient to take the effluent directly from the septic tank, the STE is synthetically prepared by using certain chemicals in their respective proportions. Table 1 shows the composition of STE and their concentration in milligram, for 1L of distilled water. The sample prepared is of $100\times$ concentrated solution and is stored at 1 °C for up to one week. The daily requirement of the sample is satisfied by suitably diluting the master sample, with tap water.

Table 1 Composition of STE

Sl. No.	Composition	Concentration (mg)
1	Peptone	160
	Meat extract	110
3	Urea	30
4	Sodium chloride	7
5	Calcium chloride	4
6	Magnesium sulphate heptahydrate	2
7	Dipotassium hydrogen phosphate	28

4 Materials

4.1 Gravel

Conventionally, several different types of media have been used for filtration. Sand or a gravel-type filter has a high porosity and permeability due to which water can flow through it, often by gravity drainage [2]. The tortuous nature of the flow path and the comparatively small pore diameters slow the flow and physically trap suspended solids. Gravel filters are most effective in reducing the turbidity of water. On the other hand, pathogens are rarely removed. Gravel sample passing through a sieve of 4.75 mm and retaining on a 2.36 mm sieve was used for filtration. The sample is washed with clear water and dried in sun.

4.2 Sand

Sand has a very important role to play in the filtration process [3]. In MDCSF, sand was used as a major filtering medium. Each drawer of the filter is filled with different grades of sand. The purity of water increases with the fineness of sand. Different types of sand with varying particle size are used in filter fabrication. This includes sand retained on 1.18 mm, 600 μm and 75 μm sieve. The drawer containing sand sample retained on 1.18 mm size is placed below gravel layer. This is followed by a drawer with 600 μm silver treated sand, which is in turn, followed by 75 μm sand.

4.3 Silver-Coated Sand

It is very well known that silver has been shown to have general antibacterial properties against a range of both Gram-negative (e.g. *Acinetobacter, Escherichia,*

Pseudomonas, Salmonella, and *Vibrio*) and Gram-positive bacteria (e.g. *Bacillus, Clostridium, Enterococcus, Listeria, Staphylococcus,* and *Streptococcus*) [4].

4.3.1 Preparation of Silver-Coated Sand

Silver-coated sand was obtained by treating sand with silver nitrate. The silver content present in purified water can resist the growth of unhealthy organisms. The steps for preparation of silver-coated sand are outlined below:

- About 500 g of graded, washed, and dried sand was mixed with 1 g silver nitrate dissolved in 1 L of distilled water, mixed thoroughly and allowed to stand for a period of 1 h.
- This mixture was then treated with 2 g of NaOH and dissolved in 50 ml distilled water and mixed thoroughly.
- The sand was treated with 10 ml of 1% of NH_4OH solution and 15 ml reducing agent (9% of sugar solution) mixed thoroughly as before and left for 1 h.
- The treated sand after solar drying was washed with distilled water to pH 7 and finally dried at 100–110 °C.

4.3.2 Activated Charcoal

Charcoal is a porous material which is often used to purify water, through the process of adsorption [5]. It is obtained by burning of wood. One of the prime reasons that activated charcoal behaves as an excellent filter material is its natural ability to remove many toxic substances from water, such as volatile organic compounds and chlorine. The steps employed for the preparation of activated charcoal are outlined below:

(a) Charcoal obtained by burning coconut shells is powdered.
(b) A 25% solution of (by weight) of calcium chloride is prepared, by weighing three parts of water and mixing in one part calcium chloride.
(c) Powdered charcoal is mixed with calcium chloride solution, and a paste is prepared.
(d) The paste is spread to dry and rinse with clean water.
(e) The paste is then baked at 225 °F for 30 min.

5 Model Fabrication

5.1 *Modified Drawer Compacted Sand Filtration (MDCSF)*

The Modified Drawer Compacted Sand Filtration (MDCSF) is a modified design for a conventional drawer compacted sand filter in which the sand layer is broken down into several layers of 10 cm height and placed in a movable drawer separated by 10 cm of air space from other layer. The new design for water treatment was based on two hypotheses: by placing the treatment media in movable drawers, separated by sufficient vertical distance, better oxygen access to the layers is facilitated, which improves the filter efficiency and facilitates maintenance requirements; the second hypothesis was that MDCSF can remove a high percentage of pollutants in STE with minimum space requirements.

This would allow such filters to be used even in locations where space is at a premium, such as dense urban areas. The comparatively lower maintenance requirements ensure that a wide range of users could easily operate the MDCSF. A laboratory-scaled model of MDCSF (Fig. 1), measuring 36 cm length × 27 cm width × 1.4 m depth with six drawers, was designed and operated. Table 2 describes the components of the MDCSF, along with their placement positions in the drawers.

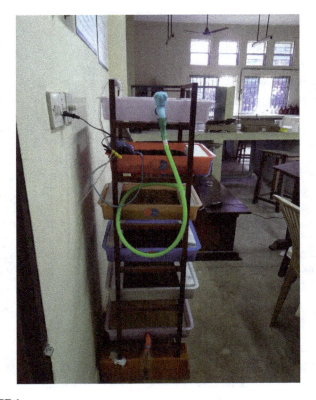

Fig. 1 MDCSF drawers

Table 2 Configuration of MDCSF

Filter media	Specifications
Drawer 1	Aeration system
Drawer 2	Gravel; effective size 4.75 mm
Drawer 3	Sand; effective size 1.18 mm
Drawer 4	Charcoal; effective size 1.18 mm to 600 μm
Drawer 5	Silver-coated sand; effective size 600 μm
Drawer 6	Sand; effective size 75 μm along with collection system
Depth of media	10 cm for each drawer, 20 cm for last one
Perforation—for each drawer—expect the first and last one	Orifice size—4 mm second drawer 2 mm—third, fourth, and fifth Orifice spacing 2 cm

6 Results and Discussions

The water quality test parameters like total dissolved solids (TDS), BOD, COD, coliform test, phosphates, nitrates, and pH were tested by standard instruments and laboratory practices, as per IS 3025(53):2003. The properties of the synthetic septic tank effluent, prepared in the laboratory, were tested first, to ensure that it behaves in the same way as the actual septic tank effluent.

Table 3 shows the comparison between the parameters of the synthetic and actual STE. Most of the parameters for the synthetic STE prepared in the laboratory were found to match with the parameters requisite for the actual STE polluted water. For the treatment of STE polluted water, about 1 L of water was fed into the first drawer. This layer was aerated by the introduction of air bubbles beneath water in the drawer.

Water from the first drawer was taken out through a pipe, into the second drawer (gravel), and uniform distribution of water to this layer was ensured using a perforated acrylic plate of 4 mm thickness. Starting from this layer, a filter paper and a perforated

Table 3 Comparison of actual and synthetic STE

Parameters	Average STE polluted water concentration	Synthetic STE concentration
Total suspended solids (mg/l)	36–85	60
BOD$_5$ (mg/l)	118–189	137.28
pH	6.4–7.8	6.9
Faecal coliform (CFU/10)	10^6–10^7	6×10^6
Total dissolved solids (mg/l)	500–30,000	867
COD (mg/l)	500–900	600
Phosphate (mg/l)	5–20	6

Table 4 Comparison of water quality parameters for treated and contaminated water

Parameters	Concentration		
	Influent	Effluent	Pure water
Total suspended solids (mg/l)	60	nil	nil
BOD_5 (mg/l)	137.28	5.3	3–5
pH	6.9	7.6	6.5–8.5
Faecal coliform (CFU/10)	6×10^6	Absent	Absent
Total dissolved solids (mg/l)	867	560	500
COD (mg/l)	600	12	10
Phosphate (mg/l)	6	2	<0.3 mg/l

acrylic plate were placed below each drawer to ensure that the filter material, along with water, does not flow into the subsequent layers. An outlet is provided in the last drawer, from where the treated water was collected and tested. A contact period of 24 h was ensured. The test results for the treated water are given in Table 4. It is observed that the straining layers in the MDCSF are effective in removing the suspended solids from the contaminated water sample.

Further, the faecal coliforms were completely removed from the sample, after treatment. Overall, it can be seen that the while the existing design of drawer compacted sand filter performs reasonably well, the modifications are explained in the study, further making it even more effective, as shown in Table 4. The minor variations in some of the water quality parameters may be due to the inherent imperfections in the model and insufficient contact periods.

7 Economic Aspects

The cost of fabrication of a prototype of a MDCSF is given in Table 5. Since the prototype is developed for small-scale filtration purposes, the actual cost of a portable working household filter shall vary. However, it is to be underlined that filtration by this system proves to be much cheaper and efficient, compared to other existing

Table 5 Cost of setting up a domestic portable MDCSF

Item	Primary cost (INR)
Stand fabrication	1500
Drawers	1500
Silver-coated sand (1 kg)	500
Pipe fittings	200
Miscellaneous	300
Total	4000

filters. A major portion of the cost is related to the initial set-up of the drawer, pipe appurtenances, and the frame. The only recurring cost is towards the cleaning of the straining sand layers, after prolonged operation. Since the drawers are movable and there is sufficient air gap between them, maintenance is easier, compared to the conventional filters.

8 Advantages and Limitations of MDCSF

The filter helps in keeping the surrounding environment (particularly, surface and ground water) from the cross contamination. Since oxygen movement is facilitated due to natural and induced aeration, no anaerobic/toxic conditions are experienced and the problems of unpleasant odours are eliminated. The maintenance operations are easy, since it involves sliding out the drawer or mixing up the media and replacing the drawer back. The entire assembly is portable; hence, it is possible to replace the filter media in any layer by a suitable material as per requirement. Since the land foot print is very less (<1.5 m^2), the filter can be placed at the rooftop, backyard, or stairwell. The filter developed is considerably cheaper compared to the available alternatives and performs extremely well, as evident from the test results. A few limitations of the MDCSF are as follows.

Due to prolonged use, silver may leach out of the silver-impregnated sand layer, which can be understood from the discolouration of the filtered water (colour changes to light brown). Also, the use of silver nitrate to prepare the silver-impregnated sand layer increases the initial cost of set-up, which is, however, easily justified in the long run.

9 Conclusions

STE is one of the important contributors of ground water pollution. In order to minimize this pollution and to have pure water for drinking, domestic as well as for industrial purposes, it is essential that STE undergoes proper treatment before it is discharged to the soil or water. Drawer Compacted Sand Filter and Vetiver grass system are the conventional techniques employed to mitigate the problem. However, the treatments results are far from satisfactory. MDCSF is a novel approach for negotiating this problem. The modifications incorporated in the design involve the use of natural/forced aeration, the use of silver-impregnated sand layer, and an activated charcoal layer. The overall results of the treatment indicate that the STE after the treatment using MDCSF meets the drinking water standards, results in increased pathogenic removal as well as other parameters like BOD, COD, TDS, *E-coli.*, etc., to remarkable extent. The aeration system provided at the initial stage reduces the bacterial growth. Also, for obtaining a better result, the antimicrobial property of silver and absorption property of charcoal were utilized by the creation of silver-impregnated sand and activated charcoal.

References

1. Mittal, A. (2011). Biological wastewater treatment. *Water Today*.
2. Jefferson, B., Palmer, A., Jeffery, P., Stuetz, R., & Judd, S. (2004). Grey water characterization and its impact on the selection and operation of technologies for urban reuse. *Journal of Water Science and Technology, 50*(2), 157–164.
3. EPA. (2002). *Onsite wastewater treatment system manual*, EPA/625/R-00/08.
4. Henry, H. (1996). Treatment of septic tank effluent using puraflo peat biofiltration system. In *Proceedings of the Eleventh Annual on-Site Wastewater Treatment Conference Minimizing Impacts, Maximizing Resource Potential*
5. Mwabi, J. K., Mamba, B. B., & Momba, M. N. (2012). Removal of *Escherichia coli* and *Fecal Coli* forms from surface water and groundwater by household water treatment devices/systems. *International Journal of Environmental Research and Public Health, 9*(1), 139–170.

Heavy Metal Fractionation in Aerobic and Anaerobic Sewage Sludge

Sooraj Garg, M. Mansoor Ahammed, and Irshad Shaikh

Abstract The study assessed the speciation of heavy metals in sewage sludge. Sewage sludge samples were collected from three full-scale sewage treatment plants which employ different treatment processes. Sewage sludge samples from activated sludge process (ASP), UASB reactor (UASBR) and moving bed bioreactor (MBBR), and one anaerobically digested activated sludge (ASP-AD) was collected during different seasons of the year. Modified sequential extraction process was used classifying metals into acid-soluble/exchangeable fraction (F1), reducible fraction (F2), oxidizable fraction (F3) and residual fraction (F4). Five heavy metals, namely chromium, copper, mercury, lead and zinc, were analysed for different fractions. Among the heavy metals, Zn (1317–1448 mg/kg) and Cu (925–1196 mg/kg) contents were the highest, followed by Cr (129–151 mg/kg), Pb (60–86 mg/kg), and Hg (18–34 mg/kg). Concentrations of all heavy metals tested except mercury in MBBR were within the limits set by different agencies.

Keywords Anaerobic sludge · Heavy metals · Sewage sludge · Sequential extraction · Speciation

1 Introduction

With increasing number of municipal wastewater treatment plants in many countries, sewage sludge treatment/management has become a particularly important problem all over the world. Among the largest producers of sewage sludge include European Union and USA producing about 10 and 6.3 million tonne/year, respectively. Daily per capita sludge production varies in the range of 0.03–0.07 kg/capita/day in developing countries [1]. Contaminants like heavy metals, organic pollutants and pathogens in wastewater are concentrated in sewage sludge through wastewater treatment process, which threaten environment and human health.

S. Garg · M. M. Ahammed · I. Shaikh (✉)
Civil Engineering Department, SV National Institute of Technology, Surat, India
e-mail: shaikhirshad1990@gmail.com

Land application is the most commonly used method all over the world for sewage sludge disposal and is being considered as one of the most economical ways for sludge disposal. This is because the sewage sludge represents a good source of nutrients such as nitrogen, phosphorus, potassium and other nutrients for agriculture reuse. However, the presence of toxic heavy metals in the sewage sludge greatly restricts its use as a fertilizer [2]. Heavy metal pollution affects the use of sewage sludge. Heavy metals such as Cu, Cd, Pb, Hg and Cr are found at relatively high concentrations in sewage sludge. The total heavy metal content of sewage sludge is about 0.5–2.0% (dry weight), and in some cases may be as high as 4% particularly for metals such as Cd, Cr, Cu, Pb, Ni and Zn [3].

The determination of total heavy metal content does not provide useful information about the risks of bioavailability, the capacity for remobilization and the behaviour of the metals in the environment [4]. It is necessary to distinguish their forms and assay their quantities. This may be achieved through speciation analysis. The most popular are chemical sequential extractions, which consist of treating a sample with chemical solutions of various leaching strength. For this, various extraction schemes (both single and sequential) were developed in the early 1980s, but most of them are modification of three-step extraction technique developed by Tessier and Rudd [5, 6].

A number of studies have been reported in the literature on the speciation of heavy metals in sewage sludge. However, very few studies compared the concentration of heavy metals in sludges from treatment plants using different treatment processes. Also, no studies have been reported from India on speciation of heavy metals in sewage sludge. The objective of this study was to assess the speciation of heavy metals in sewage sludge. Sewage sludge samples were collected from three full-scale sewage treatment plants which employ different treatment processes.

2 Materials and Methods

2.1 Sewage Sludge

For this study, sewage sludge was collected from three different treatment plants of Surat, India, which use different treatment technologies (Table 1). In Anjana sewage treatment plant, activated sludge process is used for treatment of wastewater. Sludge generated from primary settling tank and secondary settling tank is mixed and is anaerobically digested. The sludge cake after sun-drying is sold as fertiliser. In Bamroli sewage treatment plant, upflow anaerobic sludge blanket (UASB) reactor with extended aeration is used. Sewage sludge generated during the anaerobic treatment process is sent to the sludge drying bed for dewatering. In Khajod sewage treatment plant, moving bed bioreactor (MBBR) is adopted for the treatment, and the sludge generated from primary settling tank and secondary settling tank is anaerobically digested and then dried in drying bed which is sold for use in agriculture.

Table 1 Details of sludge collection points from sewage treatment plants

Wastewater treatment plant	Capacity (million litre per day)	Treatment process	Collection of sludge sample
Anjana Sewage Treatment Plant	82.50	Activated sludge process (ASP)	1. Raw activated sludge (ASP-Raw) 2. Anaerobically digested sludge (ASP-AD)
Bamroli-Vadod Sewage Treatment Plant	100	Upflow anaerobic sludge blanket (UASB) reactor with extended aeration process	1. UASB reactor sludge (UASBR)
Khajod Sewage Treatment Plant	25	Moving bed biofilm reactor (MBBR)	1. Raw MBBR sludge (MBBR)

Details of the sewage treatment plants are shown in Table 1. Sludge samples were collected monthly for four months from all the sampling points.

For analysis of physicochemical parameters, 1:10 (v/v) solution of sludge samples was mixed by a magnetic stirrer at 1000 rpm for 10 min. For extraction of heavy metals, the sludge samples were air-dried for 2–3 days and was ground in an electrical grinder and sieved through 212-μm sieve and stored at 4 °C in plastic pouches.

2.2 Sequential Extraction of Heavy Metals in Sludge

For the determination of heavy metal concentrations in sewage sludge samples, modified sequential extraction (modified BCR extraction) process was opted. During the extraction, metals were classified into acid-soluble/exchangeable fraction (F1), reducible fraction (F2), oxidizable fraction (F3) and residual fraction (F4) [7]. The detailed procedure is described as follows.

Step 1: Extraction of acid soluble/exchangeable fraction (F1): 0.5 g sludge sample was added in a 50-mL polypropylene centrifuge tube containing 20 mL of 0.11 mol/L acetic acid and was shaken for 16 h at room temperature. The solution and solid phases were separated by centrifugation at 4000 rpm for 20 min. Subsequently, the suspension was filtered through a 0.45-μm membrane filter and the solid residues were preserved for subsequent extractions.

Step 2: Reducible fraction (F2): The residues from Step 1 were shaken with a portion of 20 mL of 0.1 mol/L hydroxylammonium chloride (adjusted to pH 2 with nitric acid) for 16 h. The extraction procedure was the same as mentioned in Step 1.

Step 3: Oxidizable fraction (F3): The residues from Step 2 were dispersed in 5 mL of hydrogen peroxide (30%) and digested at room temperature for 1 h with occasional shaking. A second 5-mL aliquot of hydrogen peroxide was introduced into and digested at 85 °C (water bath) for 1 h. The contents were evaporated to a small volume (1–2 mL). 25 mL ammonium acetate (1.0 M, adjusted to pH 2 with nitric acid) was added to the cool and moist residue. The sample was then shaken, centrifuged and the extract was separated as described in Step 1.

Step 4: Residual fraction (F4): 5 mL HNO_3 was added to the residues from Step 3. The contents were heated on a hot plate and evaporated to almost dryness. After cooling, the residues were dissolved in 5% (v/v) HNO_3. The resultant solutions obtained in different steps were subsequently used to determine the heavy metals.

A blank was also run at the same time. The concentrations of Cu, Cr, Hg, Pb and Zn in different fractions and in the resultant solutions obtained in Step 4 were determined by ICP-AES. Tests on each sample were conducted in triplicate, and the average value of results is reported.

2.3 Analysis

Total solids, pH, electrical conductivity, volatile solids, total Kjeldahl nitrogen (TKN), phosphorus (P) and faecal coliforms of sludge were determined as per standard methods [8]. Heavy metals (Cu, Zn, Pb, Cr and Hg) were analysed using ICP-AES (ARCOS Spectro, Germany).

3 Results and Discussion

3.1 Characteristics of Sludge

The mean values of physicochemical properties of sludge are presented in Table 2. The results show that the mean pH of sewage sludge ranged between 7.46 and 7.68 irrespective of treatment processes. Organic matter (measured as volatile solids) and nutrients (N and P) were high in sewage sludge irrespective of treatment process. Anaerobically digested sludge (ASP-AD) had a very low concentration of organic matter and low electrical conductivity. Total solid contents of raw activated sludge (ASP-Raw) and MBBR sludge were 19.91 and 18.52 g/L, respectively, while total solid content of UASBR sludge was 11.79 g/L. Sludge samples had moderate electrical conductivity that might increase the cation exchange capacity of soil on which this sludge will be applied. It was observed that for all sludge samples, there was high TKN, indicating that these sludge samples are rich source of nutrients which make them suitable for agriculture application [9]. High concentrations of organic matter (volatile solids) may lead the formation of humic acids in soil after its application on land which may decrease the pH of soil which indicates that pH of soil should be checked periodically after and before the application of sewage sludge

Table 2 Characteristics of sludge from selected sewage treatment plants

Sample	pH	Electrical conductivity (mS/cm)	Total solids (g/L)	Phosphorus (mg/kg)	Total Kjeldahl nitrogen (mg/kg)	Faecal coliforms (MPN/g)	Volatile solids (%)
ASP-Raw	7.68 ± 0.18	2.91 ± 0.06	19.91 ± 0.80	1019 ± 59	1951 ± 124	4038 ± 508	60.27 ± 4.60
ASP-AD	7.43 ± 0.13	2.03 ± 0.09	-	772 ± 48	1589 ± 178	691 ± 191	17.1 ± 0.49
UASBR	7.46 ± 0.14	2.74 ± 0.17	11.79 ± 0.64	848 ± 60	1828 ± 129	4307 ± 279	44.35 ± 0.72
MBBR	7.56 ± 0.15	2.85 ± 0.13	18.52 ± 0.75	852 ± 31	1880 ± 89	4243 ± 501	51.6 ± 2.14

Mean ± SD based on the analysis of four samples collected at different months (triplicate for each sample)

[10]. Concentration of TKN was the highest in the anaerobically digested sludge (ASP-AD).

Microbial content is one of the most important parameters of sludge when it is used for agricultural purpose. All sewage sludge samples had very high amount of faecal coliforms exceeding the limits prescribed by USEPA except anaerobically digested sludge (ASP-AD). Table 2 shows that there was no significant variation in concentrations of different parameters in different sludge samples collected in different months.

3.2 Heavy Metal Content

Table 3 shows the variation of total heavy metals in the sludge. A comparison of metal concentration shows that there is no significant difference in the concentration of a particular metal in sludge samples collected from different treatment plants except for lead and mercury.

In general, the sludge samples had higher concentrations of Zn and Cu but relatively low concentrations of Cr, Pb and Hg. Similar results have been reported in the literature [2, 11, 12]. The amount of Zn was much higher than that of other heavy metals which could be attributed to the fact that almost all urban water supply pipes in Surat city are made of galvanized material. Cu content was also high, whereas Pb and Hg contents were relatively low. These results concur with the findings in Guangzhou, China [11]. Comparing the values of the present study with the mean values reported in India [13], the former exhibited lower concentrations for Cu and Zn but higher concentration of Pb and Hg. The results of the experiments in this research generally showed low risk of heavy metal toxicity of the sludge samples due to their low concentration with respect to the regulatory limits specified, and therefore a reasonably good potential for the sludges from the treatment plants is to be used for agricultural purposes. It may be noted that the total concentration of heavy metals in anaerobically digested sludge (ASP-AD) was higher than their concentration in raw sludge presumably due to the removal of organic matter during anaerobic digestion.

Table 3 Concentrations of heavy metals in sludge

Sample	Total concentrations of heavy metals (mg/kg)				
	Chromium[a]	Copper[b]	Lead[a]	Mercury[a]	Zinc[b]
ASP-Raw	129 ± 18.14	1011 ± 30.15	73.78 ± 12.41	18.41 ± 8.21	1387 ± 29.31
ASP-AD	151 ± 19.12	1196 ± 51.01	86.10 ± 14.21	21.37 ± 14.2	1261 ± 34.07
UASBR	132 ± 9.08	929 ± 41.66	83.62 ± 8.29	18.25 ± 12.04	1317 ± 15.95
MBBR	133 ± 8.78	925 ± 45.07	60.01 ± 11.28	34.45 ± 10.08	1448 ± 13.75

Mean ± SD based on the [a]3 samples [b]4 samples (triplicate for each sample)

3.3 Fractionation of Heavy Metals in Sewage Sludge

Fractions of Cr, Cu, Pb, Hg and Zn were determined by the sequential extraction procedure in the sludge and are presented in Fig. 1. The greatest extraction percentage was obtained in oxidizable fraction (F3) and residual fraction (F4) of all the sludge samples, which was expected given the affinity of organic matter and minerals for this type of elements and the formation of stable complexes [12]. Distribution of heavy metal fractions varied widely for different metals, as shown in Fig. 1.

Results indicate that copper (Cu) contents of acid exchangeable fraction (F1) and reducible fraction (F2) were lower than 10% in raw activated sludge (ASP-Raw) and anaerobically digested sludge (ASP-AD), thus implying less direct toxicity to the environment. Together F3 and F4 fractions accounted for 80–90% of total Cu in all

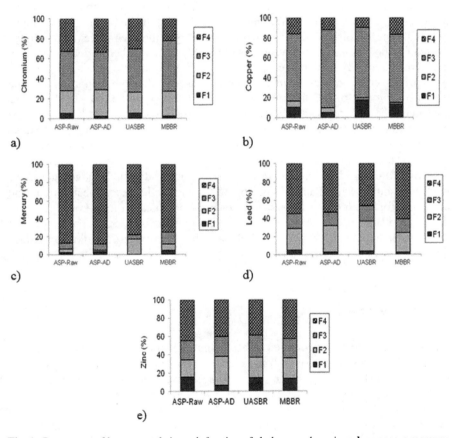

Fig. 1 Percentage of heavy metals in each fraction of sludge—**a** chromium, **b** copper, **c** mercury, **d** lead, **e** zinc. F1-acid exchangeable fraction; F2-reducible fraction; F3-oxidizable fraction; F4-residual fraction

the sludge samples, indicating that Cu was associated with strong organic ligands and probably occludes in such minerals as quartz and feldspars.

Chromium was principally distributed between F3 and F4, and these fractions accounted for over 70–75% of Cr as reported in the literature [14]. In fraction F_2, Cr did not exceed 27%, and in fraction F_1, the contents of Cr did not exceed 5%, indicating a low direct effect for the plants.

Lead was mostly present as F4 for all sludge samples. The results were in good agreement with the reports of other studies [15]. Pb can reportedly be immobilized by insoluble salts, such as phosphates [15]. The high concentration of Pb in F4 in municipal sludge indicates its low direct and potential bioavailability to the environment if the sludge is used as a fertilizer [11]. However, Pb was also present in F2 in large quantities (20–35%) that poses a threat for the application of sludge in agriculture.

Mercury was mostly present as F4 (75–88%) in all the sludge samples. The results were in good agreement with other reported studies [2]. It can predict that it is low. This indicates low direct and potential bioavailability to the environment if the sludge is used for the organic modification of soil [2]. Mercury was below detectable level in F1 fraction of UASBR sludge.

Zinc showed the greatest degree of mobility based on the high proportion of metal extracted from F1 (6–16%), F2 (11–32%), F3 (21–24%) and F4 (39–45%). Zn thus appeared to have high bioavailability and potential ecotoxicity, which is in agreement with the literature [13]. Since a part of Zn can be occluded inside crystalline structures and not readily available for plant absorption, the limits of Zn in sewage sludge are higher [11].

Among all samples (including raw and anaerobically digested sludge), the active fractions (sum of F1, F2 and F3) of Zn and Cu had the highest concentrations and accounted for over 60 and 80%, respectively. For anaerobically digested sludge (ASP-AD), the percentage of heavy metals in F1 was decreased up to 50% compared to the raw activated sludge (ASP-Raw). It indicates that the sludge digestion can reduce the possibilities of mobilization of heavy metals in sewage sludge.

4 Conclusions

The analysis of total concentration of heavy metals showed that except for mercury in MBBR, the concentrations of all selected heavy metals in all sludge samples were below the limits suggested by different agencies. Speciation of heavy metals in sludge samples showed that, while Cu and Cr were primarily distributed in oxidizable fraction, Zn, Pb and Hg were primarily distributed in residual fraction showing low direct and potential ecotoxicity to the environment. Based on the analysis of the composition of municipal sludge and permissible limits for land application, and from an environmental protection perspective, these sludge samples have good quality with respect to their application in agriculture.

References

1. Christodoulou, A., & Stamatelatou, K. (2016). Overview of legislation on sewage sludge management in developed countries worldwide. *Water Science and Technology, 73*(3), 453–462.
2. Li, J., Luo, G., Gao, J., Yuan, S., Du, J., & Wang, Z. (2015). Quantitative evaluation of potential ecological risk of heavy metals in sewage sludge from three wastewater treatment plants in the main urban area of Wuxi, China. *Chemistry and Ecology, 283*(3), 1–17.
3. Xu, G., Liu, M., & Li, G. (2013). Stabilization of heavy metals in lightweight aggregate made from sewage sludge and river sediment. *Journal of Hazardous Materials, 260,* 74–81.
4. Amir, S., Hafidi, M., Merlina, G., & Revel, J. (2005). Sequential extraction of heavy metals during composting of sewage sludge. *Chemosphere, 59,* 801–810.
5. Jamali, M. K., Kazi, T. G., Afridi, H. I., Arain, M. B., Jalbani, N., & Memon, A. R. (2007). Speciation of heavy metals in untreated domestic wastewater sludge by time saving BCR sequential extraction method. Journal of Environmental Science and Health. Part A. *Toxic/hazardous Substances and Environmental Engineering, 42*(5), 649–659.
6. Turek, M., Korolewicz, T., & Ciba, J. (2005). Removal of heavy metals from sewage sludge used as soil fertilizer. *Soil and Sediment Contamination, 14,* 143–154.
7. Wang, P., Zhang, S., Wang, C., Hou, J., Guo, P., & Lin, Z. (2008). Study of heavy metal in sewage sludge and in Chinese cabbage grown in soil amended with sewage sludge. *African Journal of Biotechnology, 7*(9), 1329–1334.
8. APHA/AWWA/WEF. (1998). In L. S. Clesceri, A. E. Greenberg, & A. D. Eaton (Eds.), *Standard Methods for the Examination of Water and Wastewater* (20th ed.). Washington, DC: American Public Health Association.
9. Mtshali, J. S., Tiruneh, A. T., & Fadiran, A. O. (2014). Characterization of sewage sludge generated from wastewater treatment plants in Swaziland in relation to agricultural uses. *Resource and Environment, 4*(4), 190–199.
10. Sastre, I., Vicente, M. A., & Lobo, M. C. (1996). Influence of the application of sewage sludges on soil microbial activity. *Bioresource Technology, 57,* 19–23.
11. Liu, J., & Sun, S. (2013). Total concentrations and different fractions of heavy metals in sewage sludge from Guangzhou, China. *Transactions of Nonferrous Metals Society of China (English Edition), 23,* 2397–2407.
12. Scancar, J., Milacic, R., Strazar, M., & Burica, O. (2000). Total metal concentrations and partitioning of Cd, Cr, Cu, Fe, Ni and Zn in sewage sludge. *Science of the Total Environment, 250,* 9–19.
13. Dubey, S. K., Yadav, R. K., Joshi, P. K., Chaturvedi, R. K., Goel, Yadav, R., & Minhas, P. S. (2006). Agricultural use of sewage sludge and water and their impact on soil water and environmental health in Haryana, India. In World Congress of Soil Science, Philadelphia. Pennsylvania, USA (Vol. 18, pp. 9–15).
14. Alvarez, E. A., Mochon, M. C., Sanchez, J. C. J., & Rodriguez, M. T. (2002). Heavy metal extractable forms in sludge from wastewater treatment plants. *Chemosphere, 47,* 765–775.
15. Walker, D. J., Clemente, R., Roig, A., & Bernal, M. P. (2003). The effects of soil amendments on heavy metal bioavailability in two contaminated Mediterranean soils. *Environmental Pollution, 122,* 303–312.

Environmental Remediation of Oil Contaminated Soil

A. Nishida, Aparna Gopinath, S. Chandraj, K. Radhika, and R. Sethu

Abstract The paper deals with study of environmental remediation of oil contaminated soil using micro-bubble generators. The method is a surfactant-free and environment-friendly method. The micro-bubble generator designed here incorporates an innovative design which will ensure greater production of micro-bubbles, and as a result, greater efficiency in the soil remediation occurs. The design concept emphasizes on vortex flow of fluids through a body with micro-holes in order to produce micro-bubble generation more effectively.

Keywords Soil remediation · Micro-bubbles (MB's) · Micro-bubble generator (MBG) · Micro-nanobubbles (MNB's)

1 Introduction

Soil contamination or soil pollution is a part of land degradation which is caused by the presence of xenobiotic chemicals or other alterations in the natural soil environment. It is typically caused by industrial activity, agricultural chemicals, or improper disposal of wastes. Oil contamination of soil is the most important threat to the

---

A. Nishida · A. Gopinath · S. Chandraj · K. Radhika (✉) · R. Sethu
Department of Civil Engineering, Muthoot Institute of Technology and Science, Varikoli, Ernakulam, India
e-mail: radhikak037@gmail.com

A. Nishida
e-mail: nishidaa@mgits.ac.in

A. Gopinath
e-mail: aparnagopinath110397@gmail.com

S. Chandraj
e-mail: chandraj123@gmail.com

R. Sethu
e-mail: mastersethu@gmail.com

natural ecosystem. Oil and oil products are the most priority pollutants of the environment because of their toxicity, spreading scale, and high migration ability [1]. Oil fields development, exploitation, and violation of the hydrocarbon transportation rules result in the pollution of natural ecosystem, particularly, soil cover. In order to find a perfect solution to the problem, it is expedient to carry out studies on individual pollutants and to adopt an effective remediation technique. Several remediation technologies based on physical, chemical, and biological methods have been employed to get rid of oil contamination of soil. However, none of these technologies score high due to one or more associated drawbacks [2]. A novel chemical-free approach for cleaning oil contaminated soil with self-collapsing air micro-bubbles with diameter less than 50 μm can be developed without the use of chemicals, such as surfactants and alkalis. Micro-bubbles have been widely explored for noteworthy applications in various fields of science and technology. They possess certain unique properties such as long stagnation time, ability to shrink, and finally, collapse under water surface while generating pressure waves in contrast to the ordinary bubbles greater than 50 μm. These properties make them suitable for cleaning applications and insight for further development of chemical-free and sustainable fine bubble-based cleaning technology for oil contaminated soil remediation is provided. MB's tend to gradually decrease in size and subsequently collapse due to long stagnation and dissolution of interior gases into the surrounding water. They have larger surface area, higher mass transfer rate, and slower rising speed than milli-bubbles. Rising velocity is a key parameter affecting the behavior of the bubbles in water and liquid solution [3]. A decrease in bubble rising velocity is associated with a decrease in bubble size. Compared to conventional bubbles in millimeter range, MB's offer novel and unique properties. These include high surface area to volume ratio, slow rising velocity in the liquid phase, and higher internal pressure. An important property of MB's that distinguishes them from conventional bubbles is that they shrink when their size is below a critical value; the rate of shrinkage significantly increases as MB size decreases, due to increased internal pressure.

2 Aim and Objectives

The primary aim is to adopt a novel chemical-free approach in the treatment of oil contaminated soil. To achieve this aim, the following objectives are set:

- To enhance the ex situ application of conventionally used soil washing method by replacing it with a surfactant-free innovative approach.
- To introduce the working application of MBG to aid for the effective remediation of oil contaminated soil.
- To enhance the production and efficiency of operation by replacing the conventional generator design by a vortex flow inducing design.
- To introduce an internal element in the generator that improves the efficiency of bubbles generated through the outlet.

3 Methodology

The study focuses on the treatment of oil contaminated soil by means of the adsorption properties of MB's generated by an innovative vortex flow MB generator. The project gets initiated with the collection of soil sample from two prominent localities, one from the near locality of Oil Refinery in Ambalamugal and the other from the flood affected area in Aluva. The sample collected from Aluva was found to be of flood washed soil having a trace amount of oil and the area gained rather much social significance due to the same. The soil samples so collected were classified as Type A and Type B soils, respectively. The sample soil was collected with different proportions of oil, and the condition of the soil before and after remediation was recorded, and the results were tabulated for further inference. Application of vortex flow inducing design in the conventional MBG provides a greater negative pressure area, thus reducing the time consumed for bubble production. Also, for greater bubble generation, an internal element is introduced within the body of the generator. The project emphasizes the large scale application of the model as a suggested alternative to soil washing, a method with similar layout of application. The design so introduced owes to be an innovative chemical-free approach to serve the cause. The project can hence be applied in areas where soil is contaminated with oil under ex situ conditions and can be used for the effective remediation of the same.

4 Principle of Design

The large scale application of the remediation method so adopted can be used as a replacement to the conventionally used soil washing method and owes to be an effective replacement in terms of reduced complexity in the process followed as well as eliminating the need for surfactants to carry out remediation. It also introduces a working design for generating MB's that serves the purpose with an indentation to its scope of being extended as a large scale design [4]. Principal of design is based on the TRIZ model, wherein the limitation posed by the conventional model vis-a-vis higher productivity and greater negative pressure resolved [5].

4.1 Conventional Design Model

Sadatomi's MNB generator is a typical bubble generating device. Its structure includes a circular pipe, nozzle, spherical body, suction chamber, micro-pore, mixing chamber, and diffusion tube. The front end of the pipeline is provided with a nozzle. The spherical body is arranged in the middle of the pipe. The suction chamber is arranged on the pipeline after the spherical body and the micro-hole is arranged on the air suction chamber pipe. The liquid from the nozzle is launched into the pipeline.

When the liquid flowing through the spherical body, the spherical body forms a negative pressure area, which is behind the sphere. Under the action of negative pressure, gas is inhaled into the mixing tube from micro-pores on the wall. In the mixing tube, gas is accelerated and dispersed by high-speed movement of the droplet and gas. Then, in the diffusion tube section, gas is compressed into MNB's. But the efficiency of MB formation is found to be low in case of Sadatomy's MNB generator, since there exists a pair of technical contradiction between "higher productivity" and "greater negative pressure." In other words, the greater the negative pressure, the more the amount of MB generated relatively and to get more negative pressure, more small spheres need to be set. But it is difficult to set more spheres in the circular pipe and to manufacture more small holes [6].

4.2 TRIZ Model

By analyzing the main problem of Sadatomi's MNB generator, there exists a pair of technical contradiction between "higher productivity" and "greater negative pressure." So, this study leads to the innovative principles recommended by TRIZ.

Working process of TRIZ model of MB generator is such that the gas–liquid mixture is firstly formed. Under the combined action of density field, centrifugal force field and negative pressure, the gas is separated from the gas–liquid mixture. Then, the gas moves toward the axis, in which process, the bubbles are gradually broken into MNB's by small spheres. Firstly, high pressure gas–liquid mixed fluid does rotary motion along the swirl spinning motion in the vortex body and is gradually accelerated along the direction of the vortex body cone. In the interaction of fluid density field and centrifugal force field, a negative pressure is generated in the center of the column at the vortex body. Gases, which are separated from the high pressure gas–liquid mixed fluid, are gradually collected to the negative pressure. While flowing through the surface of the sphere, the high pressure gas–liquid mixed fluid is continuously accelerated, and gases are continuously further cut into minute bubbles by liquid. Meanwhile, the bubbles in the negative pressure move toward the nozzle. When the bubbles are ejected from the nozzle, the high-speed strength shear and the high frequency pressure change are generated at the gas–liquid contact interface, and MNB is generated [7].

4.3 Innovative Design Model of MB Generator

The new model adopts the principle of the TRIZ model and also introduces an internal element as an innovative approach to maintain the higher productivity and greater negative pressure that the generator should maintain. The outer body casing of the design model developed of the generator is a combination of a cylinder and a frustum with a cylindrical shape at the top portion of body of the generator and toward the bottom, the shape of the frustum. Though designed as to induce vortex flow and bring in the necessary negative pressure required within the casing of the generator. The total height of the body was limited to 70 mm. From the possible ratios so derived, the heights of the cylindrical and frustum portion were fixed as 20 mm and 50 mm, respectively. The upper and lower diameters of the casing were fixed to be 60 mm and 30 mm, respectively. Also, the body of the generator was made using 3D printing technology. The internal element so designed to satisfy the need to generate sufficient MB's to carry out the remediation process effectively. As the model designed is used to meet small scale requirements and standards, the number of micro-holes introduced in the internal element is hence less. The MB's are introduced in the internal element by drilling laser holes onto the element. The number of micro-holes so introduced to serve the purpose was fixed to be 75. For this, initially stainless steel sheet of 1 mm thickness was chosen. Onto this, micro-holes were drilled. Further, the sheet was bent to suitable shape to fit into the body of the generator. The design adopted in this paper considers square-shaped internal element. Thus, the internal element was welded into square shape and further inserted into the body of the generator. Hence, the basic body of the generator was designed. Figure 1 depicts the design layout of innovative micro-bubble generator with micro-porous element.

In order to determine the working condition of the model prepared, different trials were conducted prior to undertaking the remediation process. Initially, the working efficiency of the internal element and its efficiency to yield MB's were tested. For this, the internal element, before being inserted into the casing of the generator, was taken separately and subjected to pressurized air at different ranges under standard temperature conditions provided in the laboratory. After a number of trials, the pressure to be introduced in the internal element was fixed to be 150 φ. Further, the compressor was introduced such that it aids in providing pressurized air into the internal element. The element along with the body and the air inlet together

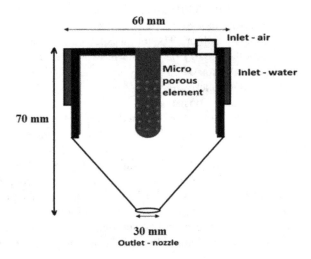

Fig. 1 Design layout of the micro-bubble generator

was sealed together to form an airtight assembly. Further, a hole was drilled onto the body of the generator at 45° inclination that serves as the water inlet into the generator. The water was introduced into the generator using a diaphragm water pump of specifications 12 V, 0.25 HP, and a flow rate of 4.5 LPM. The assembly was made working using sufficient electric power and was set for different trials to ensure proper working before remediation. Initially, the whole assembly was tested with the bubbles being dispersed into water alone. It was observed that MB's were introduced along with ordinary bubbles and the presence of the former induced a feeble lathered appearance onto the surface of water. Further, after ensuring that the model is effectively in working condition, it was taken for the purpose of remediation.

Initially, the remediation was carried out in all different samples considered, with the aid of manual stirring. For this purpose, the soil was placed in the container, and the container was rotated manually to induce stirring, and MB's were introduced into the container. The soil added with oil, when taken for remediation, was found to form clusters of soil-oil particles having different lump sizes as settled particles, suspended particles, and particles that floated on the surface of water. The remediation was carried out on the same and the time rate was noted based on the effectiveness with which the process is carried out. This was further determined as a measure of the time taken for the clusters formed due to the addition of oil to settle at the bottom surface of the container after being broken apart into individual particles having the physical appearance and characteristics as that of normal soil. The remediated soil was then taken out; oven dried. It was observed that the remediated oven dried soil was found to have changed its color of appearance from a darker shade to that comparable with normal soil which was not added with oil.

5 Design Layout

The designed model is arranged with certain assemblies such that, all such additions, together as a single unit, serve the purpose of effective remediation of soil contaminated with oil. The unit as such can be used to serve the purpose of in situ remediation. To carry out ex situ remediation, the same layout can be inferred to, with it being arranged on-site as on a temporary basis, the whole assembly mounted on a necessary platform [8].

The assembly consists of a chamber or container onto which the soil from the site, contaminated with oil is collected. From there, the soil is transported to the central chamber using conveyor belts. The central chamber is the tank of required design specifications in which the remediation process is carried out. The chamber is placed at an elevated height from the ground or site of remediation. The chamber unit consists of a mechanical stirrer or blades which serve the purpose of mixing the soil served into the chamber. The MBG is fixed onto the same chamber. For ex situ remediation, more number of MBG's, arranged in series with each other, can be incorporated onto the chamber of larger capacity and consequently convenient design parameters to carry out effective remediation. In the remediation chamber, both mechanical stirring and

MB generation take place simultaneously. The mechanical stirring is introduced so as to ensure uniform remediation throughout the soil mass. The MB's generated induces surficial adsorption of oil onto the surface of the bubbles. This activates the remediation of the oil contaminated soil. Further, the water after remediation is washed away from the chamber. For this, valves are provided to carry the post-treatment water from the chamber. The water so collected can be treated and reused [9]. As MBG's are conventionally used for water treatment, the same can be used to serve the purpose [10]. Thus, the treatment unit is provided with two water tanks, one carrying a batch of fresh or treated water and the other carrying the water after remediation. The soil after remediation is collected in a tank and laid back in the site and allowed to dry at natural temperature conditions. The MBG is considered as a potential approach in the soil treatment, though being used conventionally for water treatment purpose. So, to induce remediation, trials are conducted on the soil collected. For this, the soil is collected from two sites and is added with oil in 0, 1, 2, 3, 4, and 5% of the mass of the soil collected. Each of the soil specimens, added with oil, is made to undergo remediation. Thus, soil added with oil is collected in necessary quantity that is compatible with the design specifications of the integrated unit. Further, it is allowed to pass into the remediation chamber through conveyor belts. The soil in the remediation chamber is subjected to remediation induced by surficial adsorption of the MB's, aided by the generator, and mechanical stirring induced by the blades fixed onto the chamber. The remediation is carried out on the basis of time rate and effective remediation is inferred to with the clusters of the soil added with oil being broken down into individual soil particles that eventually settles at the bottom of the chamber. The soils after remediation are collected and the oil left in the same is quantified using Soxhlet apparatus. Figure 2 shows the design layout of the large scale remediation approach using micro-bubbles, where 1 is conveyor for separating soil, 2 is tank storing separated soil, 3 is conveyor carrying soil for mixing, 4 is mixing chamber, 5 is motor-driven stirrer, 6 is series of MBG's,

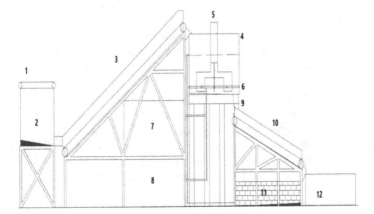

Fig. 2 Design layout of the innovative approach

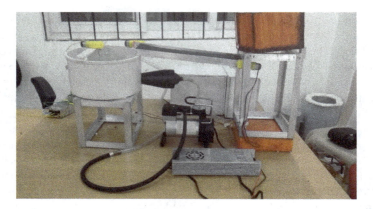

Fig. 3 Working model of the innovative approach

7 is air compressor, 8 is water filtering unit, 9 is soil outlet, 10 is conveyor carrying remediated soil with provisions of removing water, 11 is water collection tank, and 12 is tank collecting remediated soil (Fig. 3).

6 Results and Discussions

6.1 Experimental Investigations

Oil remediation of soil under an innovative approach owes to the development of a technology that reduces the complexities in functioning posed by the convention al method being currently undertaken. Prior to carrying out the remediation process, the geotechnical properties of the land have to be studied for inferring it to various civil engineering practices after the process of remediation. For this, basic geotechnical properties of soil were carried out. The comparison of the same with that of the soil added with different proportions of oil was also made. This was conducted to determine the variation of the properties of soil with the addition of oil under various proportions.

It was found that the soil tends to exhibit well-graded characteristics irrespective of whether oil was added or not, and that too unaffected by the proportion of the oil in soil. Water content, specific gravity, and strength properties of soil were found to show a decreasing trend with the increase in the proportion of oil added to soil. As the trend so obtained proves to impart an adverse effect of the civil engineering works that could be undertaken, remediation of the same becomes necessary.

6.2 Results of Remediation

The soil collected from two sites named Type A and Type B, respectively, were added with different proportions of oil at 0, 1, 2, 3, 4, and 5% of the weight of the soil. Further, the soil was treated using micro-bubble generator. The treatment was stopped at a point where the soil added with oil was free from clusters formed due to the soil–oil interaction and settles at the bottom of the remediation chamber as individual particles, free from oil. Also, the soil so treated was further oven dried and the remediation was justified with the soil added with oil having physical appearance comparable to that of the normal soil as shown in Figs. 4 and 5. The quantification of post-remediated soil is done using Soxhlet apparatus. It was inferred from the process of remediation undertaken using the innovative approach introduced to the micro-bubble generator that the percentage of remediation induced in Type A soil was greater than Type B soil at comparable time rates. Also, the process reduces the complexities in the steps involved in that of conventionally used soil washing method. The process also owes to be eco-friendly in nature as it exempts the use of chemicals throughout the steps involved (Tables 1 and 2).

Fig. 4 Physical appearance of soil Type A **a** before remediation, **b** after remediation, **c** normal soil

Fig. 5 Physical appearance of soil Type B **a** before remediation, **b** after remediation, **c** normal soil

Table 1 Remediation results of soil Type A

C_0	C_t	t in min
0.01	0.31	12.9
0.02	0.49	17.07
0.03	0.87	23.23
0.04	1.23	27.77
0.05	1.73	30.38

Table 2 Remediation results of soil Type B

C_0	C_t	t in min
0.01	0.18	12.18
0.02	0.39	15.82
0.03	0.71	20.03
0.04	1.07	27.32
0.05	1.43	32.88

6.3 Validation of Mathematical Model

MB's proved to be a promising alternative in the case of removal of oil from the soil. MBG's produce these bubbles, thus enabling the remediation of a large volume of oil contaminated soil. In order to determine the efficiency of the model, mathematical modeling can be done. Experimental breakthrough curves can be explained by using such models. Most of the developed models require a preliminary determination of large number of parameters, which require additional experimentation and nonlinear curve fitting. This mathematical complexity or the need to know too many parameters from different experiments make these models rather inconvenient for practical use. For that reason, various mathematical empirical models have been developed to predict the amount of pollutants or oil content remaining in the soil after remediation.

Bohart–Adams model is used as the basis of mathematical modeling. Bohart and Adams proposed an equation for the design of adsorption column. Considering the MB's as a good adsorbent of oil, the amount of oil removed from soil or adsorbed on to the bubbles can be determined by using this model. The model assumes that the adsorption rate is proportional to both the residual capacity of the pollutant, i.e., oil, and the concentration of the adsorbents, i.e., the MB's. The empirical formula given by the model is:

$$\ln\left(\frac{C_t}{C_0}\right) = K_{AB}.C_0.t - K_{AB}.N_0.\frac{Z}{F}$$

where C_t = effluent solute concentration (mL/g), C_0 is initial solute concentration, K_{AB} is kinetic constant (mL/g min), N_0 is saturation concentration, Z is bed depth (cm), F is linear flow rate (mL/min), and t is time (min).

Here, the modified concept is to generalize the application of the equation to any undertaken sorbent or sorbing species. Here, the influent concentration is the concentration of oil in the soil before remediation. Similarly, the effluent concentration is the concentration of oil in the soil after remediation. This is quantified using Soxhlet apparatus. K_{AB} and N_0 are obtained from the intercept and slope of the plot $\ln\left(\frac{C_t}{C_0}\right)$ versus t. Z is the bed depth which is assumed to be the depth at which the MB's are released onto the soil with respect to the surface of the soil bed which is fixed to be 1 cm. F is the linear flow rate which is assumed to be a constant value and is adjusted using valves connected to the assembly; the value so obtained is 10 mL/min. t is the time taken for remediation and is fixed to be at a rate of 10 min per cycle of remediation on a single soil specimen. The values obtained using the parameters are tabulated and graphs are plotted. The modified approach adopted to the modeling is justified using the regression value. Regression value or R^2 is a statistical measure of how close the data are to the fitted regression line. It is also known as the coefficient of determination, or the coefficient of multiple determination for multiple regression. If the value of R^2 so obtained lies between 0.83 and 0.98, then the mathematical modeling adopted can be effectively applied to all conditions concerning the assumption (Fig. 6; Tables 3 and 4).

The regression value can be expressed as:

$$R^2 = \frac{\text{sum of squares of resisdual}}{\text{total sum of squares}}$$

The regression value hence obtained from the graph for Type B soils is: $R^2 = 0.9727$.

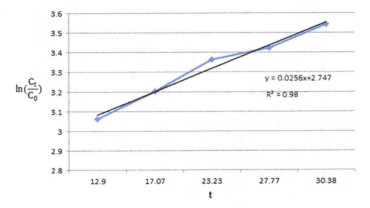

Fig. 6 Graph for Type A soil

Table 3 Mathematical modeling of Type A soil

C_0	C_t	Z	F	t
0.01	0.31	1	10	12.9
0.02	0.49	1	10	17.07
0.03	0.87	1	10	23.23
0.04	1.23	1	10	27.77
0.05	1.73	1	10	30.38

Table 4 Results obtained from graph for soil Type A

K_{AB}	N_0
0.0256	2.748

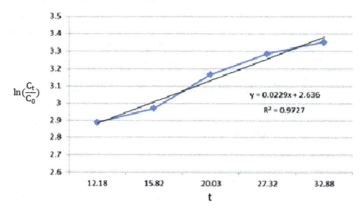

Fig. 7 Graph for Type B soil

Table 5 Mathematical modeling of Type B soil

C_0	C_t	Z	F	t
0.01	0.18	1	10	12.18
0.02	0.39	1	10	15.82
0.03	0.71	1	10	20.03
0.04	1.07	1	10	27.32
0.05	1.43	1	10	32.88

Table 6 Results obtained from graph for soil Type B

K_{AB}	N_0
0.0229	0.636

As the value of R^2 so obtained lies between 0.83 and 0.98, the mathematical modeling adopted can be effectively applied to all conditions concerning the assumption.

7 Conclusions

Oil contaminated soil has been widely recognized to constitute a major environmental issue due to its adverse effects on human health and ecological safety. There are many conventionally existing methods to treat oil contaminated soil. Of these, the commonly used technique is soil washing. But soil washing poses certain difficulties due to the complexities in the steps undertaken for remediation. Also, the method makes use of chemicals to treat soil and hence is non-eco-friendly in nature. So as to overcome such disadvantages, an innovative approach was brought about for remediation that replaces the existing method. Soil remediation using micro-bubble technology proved to be a promising alternative because of the noteworthy applications of micro-bubbles in science and technology. For the production of micro-bubbles, micro-bubble generator is designed. The generator is provided with a vortex flow casing and an internal element with laser holes drilled into it at micro-scale, which maintains high productivity with greater negative pressure. Another advantage of this method of remediation over conventional methods is that, it is a chemical-free approach and is less complex as compared to other methods. Hence, the method is eco-friendly in nature. The study also introduces a design layout that can be applied both in in situ as well as ex situ conditions, contradictory to the existing methods, which is applicable only in either of the two. Furthermore, the layout serves to have dual purpose as the water used is being treated, and thereby causing a significant reduction in the amount of water being used for remediation. Thus, the project owes to be innovative and efficient in its design, productivity, and efficiency.

Acknowledgements The authors are sincerely grateful to everyone who has supported to complete the project. Last but not the least, authors are very thankful to God Almighty.

References

1. Agarwal, A., Zhou, Y., & Liu, Y. (2016, December). Remediation of oil-contaminated sand with self-collapsing air micro bubbles. *Environmental Science Pollution, 2016*.
2. Agarwal, A., & Ng, W. J. (2011). Y Liu (2009), Principle and applications of M Band nano bubble technology for water treatment. *Chemosphere, 84*(9), 1175–1180.
3. Cheng, C. H., Yang, A. S., & Cheng, H. Y. (2017). Design and fabrication of MB oxy generator. *International Journal of Mechanical and Mechatronics Engineering, 11*(7).
4. Kim, H. S., Lim, J.-Y., Park, S.-Y., & Kim J.-H. (2018). Effects of distance of breaker disk on performance of ejector Type MB generator. *KSCE Journal*.

5. Jenkins, K. B. (1993). Application of oxygen micro bubbles for in situ biodegradation of P-xylene-contaminated groundwater in a soil column. *Biotechnology Progress, 1993*.
6. Sadatomi, M., Kawahara, A., Kano, K., & Ohtomo, A. (2016). *Performance of a new microbubble generator with a spherical body in a flowing water tube*. Department of Mechanical Engineering and Materials Science, Kumamoto University, Kumamoto 860-8555, Japan b Applied Electronics Research Center, Kumamoto Technology and Industry, 861-2202, Japan.
7. Khantia, S., Majumdar, S. K., & Ghosh P. (2012). MB aided water and wastewater purification: A review. *Reviews in Chemical Engineering, 2012*.
8. Marui, T. (2016). An introduction to micro/nano bubbles and their applications. *Systemics, Cybernetics and Informatics, 2*(4).
9. Budhijanto, W., Darlianto, D., Pradana, Y. S., Hartono, M. (2016) Application of MB generator as low cost and high efficient aerator for sustainable fresh water fish farming. In *International Seminar on Fundamental and Application of Chemical Engineering* 2016.
10. Chen, Y. (2016). Innovative design for vortex micro nano bubble generator based on TRIZ. In *3rd International Conference on Mechatronics And Information Technology*, 2016.

Treatment and Reuse of Periyar Sedimented Soil Using Nanochemicals

B. Diya and Ann Mary Mathew

Abstract Severe floods affected the South Indian state of Kerala on 8 August 2018. The overflow of Periyar River resulted in large losses during and after the flood. Aftereffects of flood were severe. There were no available land or proper method to dispose the sediment dredged waste. Also it caused a large destruction to the pavements in the surrounding area of the river due to overflow. Pot holes developed due to the ingression of water into the subgrades. It demands huge amount of money for the repair of these pavements as well as to dispose the waste collected from various buildings. So in order to utilise the soil effectively and economically in an ecologically beneficial manner, we introduce a method for the proper treatment and reuse of these sedimented wastes as a suitable subgrade material. The soil treated with the nanochemicals can be used as a subgrade for pavement construction. The nanochemicals used were Zycobond and terrasil from Zydex Industries. These chemicals were added in various percentages such as 0.02, 0.04, 0.06, 0.07, 0.08, 0.09 and 0.1% to obtain optimum dosage.

Keywords Nanochemical · Subgrade · Flood

1 Introduction

From August 2018, severe floods affected the South Indian state of Kerala, due to unusual high rainfall during the monsoon season. Chengannoor, Pandanad, Edanad, Aranmula, Kozhencherry, Ayiroor Ranni, Pandalam, Kuttanad, Malappuram, Aluva, Chalakudy, Thiruvalla, Eraviperoor, Vallamkulam, North Paravur, Vypin Island and Palakkad Chellanam are the most affected areas which had been witnessed. The aftereffects of these floods became more crucial in the case of proper disposal of the

---

B. Diya (✉) · A. M. Mathew
Department of Civil Engineering, A P J Abdul Kalam Technological University, Trivandrum, India
e-mail: diyamcity@gmail.com

A. M. Mathew
e-mail: annmarymathew92@gmail.com

© The Editor(s) (if applicable) and The Author(s), under exclusive license to Springer Nature Singapore Pte Ltd. 2021
J. Thomas et al. (eds.), *Current Trends in Civil Engineering*, Lecture Notes in Civil Engineering 104, https://doi.org/10.1007/978-981-15-8151-9_4

Fig. 1 Waste soil deposited on the banks

waste from the building which was deposited as a result of overflow of Periyar River as in Fig. 1. Cleaning of flood-hit houses, shops and flats was a very big challenge [1]. The cleaned wastes from various buildings were deposited in front of the houses, but could not find an effective solution for disposal of this soil waste in a proper manner [2]. This study indicates a method for the proper reuse of these sedimented wastes in a suitable manner using nanochemicals [3].

The main objective of the study is to take away the waste which is removed from buildings in the flood affected areas. But through this study, it is determined to utilise this waste soil as an effective and economical subgrade material. The principal aim is to protect the environment and enhance public health, while optimising the cost. Use of such materials typically results in considerable cost savings. However, such material is often very inexpensive [4].

2 Materials

2.1 Soil

Soil was collected from the front yard of a flat named Jewel homes in Aluva as in Fig. 1. 65 kg of soil was collected after digging up to 3 m. A sample soil of 500 g was collected in a zip bag for the determination of water content of the soil.

The properties of the soil are shown in Table 1.

Table 1 Properties of soil

Properties	Values
Natural water content (%)	40
Specific gravity	2.66
Clay percentage (%)	32.34
Silt percentage (%)	46
Sand percentage (%)	21.66
Liquid limit (%)	45
Plastic limit (%)	28
Shrinkage limit (%)	18.9
Plasticity index (%)	17
Optimum moisture content (%)	28.5
Maximum dry density (g/cc)	1.485
Soil classification	CI
Free swell index (%)	16.66
Organic content (%)	0
UCC strength (kg/cm^2)	0.6135
Shear strength (kg/cm^2)	0.3067
Permeability (m/s)	4.41×10^{-5}
PH	5.25

2.2 Terrasil

Terrasil is a soil modifier that permanently eliminates infiltration of water into soils [5]. Making soil bases impervious to water has a significant impact on overall life cycle costs. Terrasil was collected from Zydex Industries as shown in Fig. 2. The properties are shown in Table 2.

2.3 Zycobond

It imparts water proofing and resists water ingress through the unpaved areas [6]. 1 kg bottle as in Fig. 3 is purchased from Zydex Industries. It is mixed and diluted with water and used [7].

The physical and chemical properties of Zycobond are as shown in Table 3.

Fig. 2 Terrasil from Zydex Industries, Gujarat

Table 2 Properties of terrasil

Form	Liquid
Colour	Pale yellow
Flash point	>80 °C
Explosion hazard	Not known
Density	1.01 g/ml
Freezing point	5 °C
Solubility	Miscible with water
pH value	10% solution in water. Neutral or slightly acidic
Viscosity	100–500 CPS

Source Zydex Industries

3 Result and Discussion

3.1 Variation of Unconfined Compressive Strength (UCC) on Addition of Nanochemicals Upon Curing

Unconfined compressive strength test was done to determine the threshold ratio of nanochemical soil mix. However, nanochemicals are required to add in a small percentage. The nanochemicals were added in percentages of 0.07, 0.08, 0.09 and 0.1%. The following results were obtained and shown in Chart 1.

Fig. 3 Zycobond from Zydex Industries

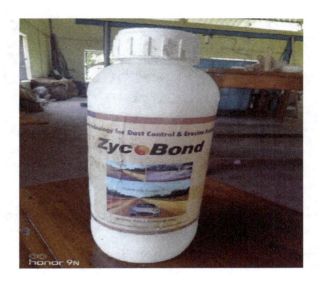

Table 3 Physical and chemical properties of Zycobond

Appearance	White liquid
Odour	Mild
Chemical type	Acrylic copolymer
Physical state	Liquid
Solubility	Dispersible in water
Density	1.01–1.02 g/ml
pH value	Approx 6.5–8
Chemical stability	Stable under normal temperature and pressure
Condition to avoid incompatibilities with other material	Strong oxidising agent, acid, alkali
Hazardous polymerisation	Hazardous polymerisation will not occur

Source Zydex Industries

4 Discussion

The UCC strength is maximum for 0.07% of nanochemical upon 3 days of curing. The value of strength decreased with further addition of the chemical as well as days of curing.

Chart 1 Variation of strength with various percentages of nanochemicals

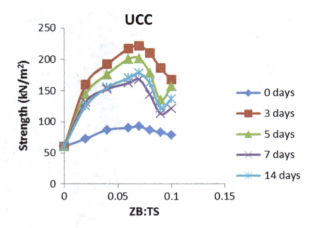

4.1 Proctor Compaction Results

Proctor compaction tests are done with the same mix chosen for UCC tests. The variation of maximum dry density is shown in Chart 2.

5 Discussion

Maximum dry density decreased as the percentage of addition of chemical increased. The maximum dry density value at $0.07\% = 1.85$ g/cc. The variation for 0.08 and 0.09% is almost equal. Maximum dry density increased by 33.4%.

Chart 2 Variation of maximum dry density (MDD) with % of nanochemicals

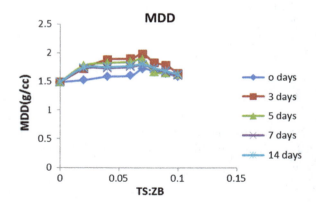

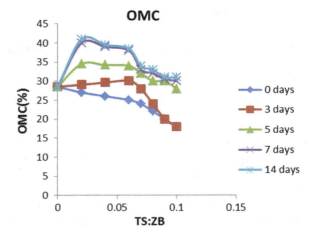

Chart 3 Variation of optimum moisture content

5.1 Effect of Optimum Moisture Content by Addition of Nanochemical

The variation of optimum moisture content with nanochemical is shown in Chart 3.

5.2 Variation of CBR

Soaked CBR value is found at different percentages of nanochemicals as in Chart 4.

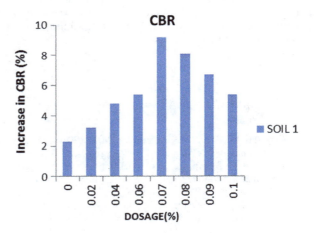

Chart 4 Soaked CBR variation

6 Discussion

The CBR value increased from 2.27 to 9.2% at 0.08% of nanochemicals can be effectively used for subgrade construction.

7 Conclusion

The objective of the study was to investigate the reuse of the wasted sedimented soil as a result of flood with the addition of nanochemical, thereby effectively utilising the soil additive mix for the construction of subgrade. The main focus was on determining the optimum ratio of the nanochemical soil mix. For that the maximum dry density, optimum moisture content, strength and CBR values are determined.

From the test results

- The OMC decreased and maximum dry density increased.
- CBR value increased to 305.28% at 0.08% nanochemical.
- The unconfined strength of the soil increased at 0.07% of nanochemical.
- The ratio of nanochemical flyash mix that can be effectively used as the subgrade material was determined as 0.07:2 with a curing period of 3 days.

Through the results in the paper, it can be concluded that 0.07:0.07:2 terrasil–Zycobond mix could be effectively used as a subgrade material. Due to the high CBR value, the thickness of subgrade can be reduced.

Acknowledgements This research was supported by providing nanochemical by Zydex Industries, Gujarat. The people from Aluva helped in collecting the soil samples. Guidance and support are given by Twinkle Vinu Mohandas who worked as an assistant professor at Marian College of Engineering. I am also grateful to my guide Mrs Ann Mary Mathew and mentor Mrs Rani V of Civil engineering and also thankful to our department head Dr. Narayanan who supported and corrected us in our technical vocabularies.

References

1. Gayathri, M., Singh, P. N., & Prasanth M. G. (2016). soil stabilization using terrasil, cement and flyash. *i manager's Journal on Civil Engineering, 6*(4).
2. Olaniyan, O. S., Ajileye, V. O. (2018). Strength characteristics of lateritic soil stabilized with terrasil and zycobond nanno chemicals. *International Academy of Engineering and Technology Science, 7*(2). ISSN (P): 2278-9987; ISSN (E): 2278-9995.
3. Taranga Divija Kalyani, V., Begum, A. S., Prasad, D. S. V., & Prasada Raju, G. V. R. (2018). A study on geotechnical properties of expansive soil treated with rice husk ash and terrasil. *The International Journal of Engineering and Science (IJES), 7*(8), 93–99. Ver. III.
4. Mulla, A. A., & Guptha, K. G. (2019). *Comparative Study and Laboratory Investigation of Soil Stabilization Using Terrasil and Zycobond* (pp. 757–769). Springer Nature Singapore Pte Lt.

5. Rohith, M. S. R., Kumaraswamy, N., Srinivasa Kumar, R., & Koteswar Rao, P. V. S. (2018a). Economic analysis of flexible pavement by using subgrade soil stabilised with zycobond and terrasil. *International Journal of Creative Research Thoughts (IJCRT)* 313–317.
6. Raghavendra, T., Rohini, B., Divya, G., Abdul Sharooq, S., & Kalyanbabu, B. (2018). Stabilization of black cotton soil using terrasil and zycobond. *International Journal of Creative Research Thoughts (IJCRT)*, 300–303.
7. Rohith, M. S. R., Srinivasa Kumar, R., Paul, W., & Kumaraswamy, N. (2018b). A study on the effect of stabilizers (zycobond & terrasil) on strength of subgrade on Bearing capacity of soil. *Indian Journal of Scientific Research, 17*(2), 86–92.
8. Pandagre, A. K., & Rawat, A. (2016). Improvement of soil properties using nonchemical—Terrasil: A review. *International Journal of Research, 3*(19).

A Study on Red Soil to Form an Bouncy Cricket Pitch

S. Amritha and V. Rani

Abstract The characteristic of soil for pitches varies considerably from country to country, and generally, pitches in Indian subcontinents are regarded to be slow and dusty in contrast to pitches in other countries. This study involves research on the behaviour of soil for cricket pitches in Kerala according to BCCI guidelines expressed in terms of percentages. The main purpose is to make use of locally available soils in order to improve its strength thereby developing a fast and bouncy pitch. For this purpose, different percentage of bentonite clay is added to the available soil in order to yield hardness to the pitch. The clayey property of bentonite clay makes the soil stiffer. Using proper clay content and appropriate techniques and proper maintenance, a perfect pitch can be developed here in Kerala. All the steps and available conditions during making process like soil selection, laboratory tests like compaction, CBR test, specific gravity, sieve analysis, etc., schedules to attain maximum compaction, CBR values and properties of soils do have proved scientific reasons and have the direct correlation with outcome and performances of pitch. This project aims on improving the characteristics of soil for an efficient rebounding pitch.

Keywords Shear strength · Compaction · California bearing ratio

1 Introduction

Cricket is considered as one of the most popular sports in the world. Playing surface, weather, ground conditions and many other variables play a part. Especially important among these is the playing surface known as cricket pitch. Here a model pitch is generated to observe characteristics of soil such as improving its strength, durability for forming an efficient pitch. A relation is generated between compaction and vertical

---

S. Amritha (✉) · V. Rani
Department of Civil Engineering, Marian Engineering College, Thiruvananthapuram, India
e-mail: amrithasunnysuni@gmail.com

V. Rani
e-mail: rani_vinoo@yahoo.com

© The Editor(s) (if applicable) and The Author(s), under exclusive license to Springer Nature Singapore Pte Ltd. 2021
J. Thomas et al. (eds.), *Current Trends in Civil Engineering*,
Lecture Notes in Civil Engineering 104, https://doi.org/10.1007/978-981-15-8151-9_5

bounce test. A ball is allowed to fall under gravity on model pitch from a height of about 2 m in order to measure the bouncing height made by the ball. The bouncing height is measured by Engauge Digitizer recommended by BCCI.

Different percentages of bentonite clay are added to the available soils to improve its physical characteristics. The liquid consistency state plays an important role in pitch preparation. Transformation between solid and plastic consistency is used in pitch preparation. The model pitch is wetted to make it plastic so it gets smoothened and then allowed to dry so it moves to solid consistency where it becomes hard as well as bouncy character gets improved.

Shannon [1] made a basic guide on pitch preparation for cricket on soils in order to make a fast and bouncy pitch. He mentioned guidelines for developing pitches on soils and also renovation of old pitches for better performances. He also suggested that the process of compaction and closely mown turf could develop perfect pitches on ground.

Parsons [2] studied the effect on aeration of clay soils in cricket pitches. In this study, effect of aeration processes on soils was analysed so he came to a conclusion that the aeration process can change physical properties and biological health of soil-based sports surfaces. Thus, he proposed guidelines for conducting aeration treatment on soil pitches, thereby providing best solutions to overcome ground problems.

Nawagamuwa et al. [3] made a study on the improvement of local soils in order to make fast and bouncy cricket pitches. In this study, he mainly focussed on the possibilities of improving characteristic of soils for producing fast and bouncy pitches by focussing on different soil properties. He concluded that from the tests conducted, the results showed that the plasticity characteristic of soil improved by the introduction of bentonite clay on the local soils.

Singh [4] made a study on cricket pitches—science behind the art of pitch making. In this study, he mainly focused on different methods to develop a perfect bouncy pitch by conducting different experiments at different percentages. And he came to a conclusion that soils with some desirable amount of clay content are considered as a perfect pitch soil.

James et al. [5] made a study on predicting the playing character of cricket pitches. In this study, he determined that the soil properties and a correlations were drawn between pitch performance and soil compaction. He concluded by finding different properties of the soil, and the amount of percentage of clay content required for an efficient bouncy pitch was also determined.

Haake [6] made a study on the playing performance of countries cricket pitches. In this study, he determined the different soil properties by conducting different experiments and conditions required for producing a fast and bouncy pitch. He concluded by finding different soil properties, and the amount of bentonite clay required to produce a fast and bouncy pitch was identified.

Table 1 Properties of bentonite clay

Properties	Value
Specific gravity (IS 2720 part 3)	2.57
Liquid limit (IS 2720 part 5)	336%
Plastic limit (IS 2720 part 5)	47%
Shrinkage limit (IS 2720 part 6)	12.4%
Plasticity index (IS 2720 part 5)	289%
Optimum moisture content (IS 2720 part 7)	40%
Maximum dry density (IS 2720 part 7) g/cc	1.19
Soil classification	CH
Free swell index (IS 2720 part 40)	120%
Undrained shear strength (IS 2720 part 10) kN/m^2	112.7
Coefficient of permeability (IS 2720 part 17) m/s	3.2×10^{-10}

2 Materials

2.1 Bentonite

Bentonite is a form of clay which comprises montmorillonite. Bentonite used in this study mainly comprises sodium ions as their major constituent. The material was collected from English India Private Ltd., Veli. A clayey material which enhances the properties of soil by its addition in varying percentages proves an efficient way in increasing the strength parameters of the soil. The bentonite property is mainly exploited to produce green moulding sand. In this application, bentonite with a suitable moisture content covers quartz sand grains and acts as a connective tissue to the entire mass. Under this homogenous coating, even at maximum compression, water will remain in a highly "rigid" state, binding the sand grains and lending maximum resistance to the sand mould. Bentonite vitrification temperature is higher than other clays. Therefore, when used as an additive, it makes green sand more durable, and, in particular, more resistant to heat stress. Table 1 shows the properties of sodium bentonite used in the study (Fig. 1).

2.2 Red Soil

Red soil is a type of soil that develops in a warm, temperate, moist climate under deciduous or mixed forest, having thin organic and organic–mineral layers overlying a yellowish-brown leached layer resting on an illuvium red layer. Red soils are generally derived from crystalline rock. They are usually poor growing soils, low in nutrients and humus and difficult to be cultivated because of its low water holding

Fig. 1 Bentonite clay

Fig. 2 Red soil. *Source* Wikipedia

capacity. The soil sample was thoroughly oven-dried, weighted and stored in sacks at room temperature (Fig. 2; Table 2).

3 Methodology

Laboratory tests were conducted to determine the engineering properties and strength characteristics of soil samples with and without addition of bentonite. The main materials characterized in the present study are red soil, bentonite. A brief introduction about these materials and methodology is explained in this chapter. The soil considered for this study was tested first for engineering properties, and the samples were tested for determination of strength parameter that is compaction. Tests were conducted on varying percentages of bentonite to both the samples—0%, 5%, 10%, 15%, 20% and 25% by weight, respectively, and optimum is found out.

Table 2 Properties of red soil

Properties	Result
Liquid limit (%)	35
Plastic limit (%)	24.47
Shrinkage limit (%)	16
Soil classification	CI
Clay (%)	57
Silt (%)	24
Sand (%)	19
Specific gravity	2.62
ucc strength (kN/m^2)	33.23
Optimum moisture (%)	17
Maximum dry density (kN/m^3)	1.7

4 Results and Discussions

4.1 Liquid Limit

The experiments were conducted for red soil with varying percentages of bentonite from 0, 5, 10, 15, 20 and 25% of bentonite along with the red soil. The variation of liquid limit for red soil with varying percentages of bentonite is shown below (Fig. 3).

There is increase in the liquid limit value with increase in percentage of bentonite clay content. The liquid limit increased up to 25% (Nawagamuwa 2014). The values obtained are within the limits and are considered suitable for remaining experiments.

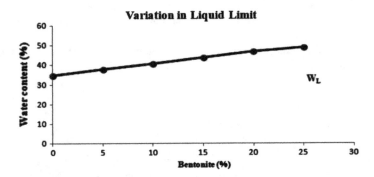

Fig. 3 Variation in liquid limit for red soil with varying percentages of bentonite

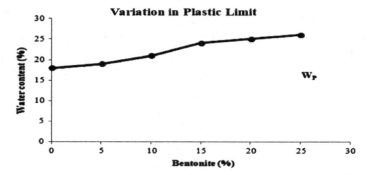

Fig. 4 Variation in plastic limit for red soil with varying percentages of bentonite

4.2 Plastic Limit

The experiments were conducted for red soil with varying percentages of bentonite from 0, 5, 10, 15, 20 and 25% of bentonite along with the red soil. The variation of plastic limit for red soil with varying percentages of bentonite is shown below (Fig. 4).

Red soil showed an increase in plastic limit with increase in percentage of bentonite clay content. The values obtained are within the limits and are considered suitable for remaining experiments (Nawagamuwa 2014).

4.3 Plasticity Index

The experiments were conducted for red soil with varying percentages of bentonite from 0, 5, 10, 15, 20 and 25% of bentonite along with the red soil. The variation of plasticity index for red soil with varying percentages of bentonite is shown below (Fig. 5).

Red soil showed an increase in the plasticity index with increase in percentage of bentonite clay content.

4.4 Compaction Characteristics

The experiments were conducted for red soil with varying percentages of bentonite from 0, 5, 10, 15, 20 and 25% of bentonite along with the red soil. The variation of compaction for red soil with varying percentages of bentonite is shown below (Fig. 6).

A Study on Red Soil to Form an Bouncy Cricket Pitch

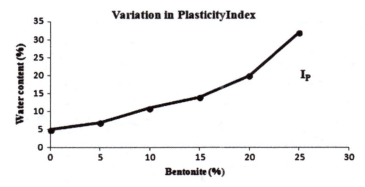

Fig. 5 Variation in plasticity index for red soil with varying percentages of bentonite

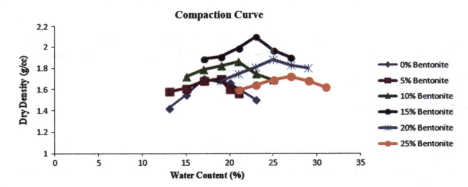

Fig. 6 Compaction curve for red soil with varying percentages of bentonite

The compaction value of red soil also increased with increase in percentage of bentonite clay content as per the journal of Usman et al. [7]. The maximum value obtained for compaction was for 15% of clay content with OMC 23% and MDD 2.1%.

4.5 Image Processing of Vertical Bounce Test

The experiment was conducted by allowing the cricket ball to fall freely from a height of about 2 m into the model pitch. By image processing, the rebound height of ball is noted and it is found that up to 15% of bentonite content bounce value increased and then it decreased. The variation of bounce height for red soil with varying percentages of bentonite is shown below (Fig. 7).

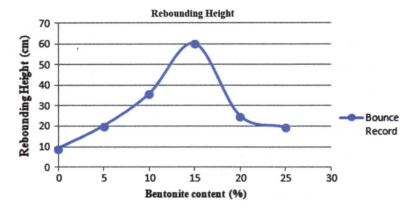

Fig. 7 Variation of vertical bounce with varying percentages of bentonite clay

The red soil showed an increase in height up to 15% of bentonite content, and then, it decreased. The optimum value of bounce is found at 15% clay content.

5 Conclusions

Following are the conclusions obtained from the study

- The compaction curve was found with increase in percentage of bentonite in red soil.
- For the optimum percentage of bentonite, the vertical bounce increased to 5.5 times than that of original state. The optimum value for bounce was observed at 15% of clay content for red soil.
- The liquid limit value was determined for red soil with varying percentages of bentonite clay from 0% to 25%. The result showed an increase in value of liquid limit with increase in percentage of bentonite clay. The value obtained for each percentage was within the limits that is between 22.5 and 49.6%, which indicates that the soil can be utilized for the making of pitches.
- The plastic limit value was determined for red soil with varying percentages of bentonite clay from 0 to 25%. The result showed an increase in value of plastic limit with increase in percentage of bentonite clay.
- The value obtained for each percentage was within the limits, which is between 13.8 and 34%. It indicates the binding effect of bentonite on soil. The compaction values were also determined for red soil with varying percentages of bentonite clay from 0 to 25%. The maximum dry density (MDD) and OMC were obtained at 15% of bentonite clay content on red soil, and for this optimum percentage of bentonite, the vertical bounce test was conducted for red soil to found its rebounding height.

References

1. Shannon, J. (2010). Basic guide for cricket pitch preparation on soils. *International Journal of Sports Engineering and Research, 8*(8), 1–56.
2. Parsons, S. (2012). The aeration of clay soils in cricket. *Journal of Geotechnical Engineering, 6*(10), 437–462.
3. Nawagamuwa, U. P., Senanyake, A. I. M. J., & Sanjeewa, D. M. I. (2014). Improvement of local soils in order to make fast and bouncy cricket pitches. *International Journal of Science and Research, 5*(5). 260-192-924.
4. Singh, S. B. (2014, July). Cricket pitches—Science behind the art of pitch making. *International Journal of Science and Research (IJSR), 3*(7).
5. James, D. M., Carre, M. J., & Haake. S. J. (2015). Predicting the playing character of cricket pitches. *International Journal of Science and Research, 8*(8), 193–207.
6. Haake, S. J. (2015). The playing performance of countries cricket pitches. *International Journal of Sports Engineering and Research, 7*, 1–14.
7. Usman, H., Hamza, M. M., Hamid, P. M., & Ahmad, T. (2016). Improvement of geotechnical properties of cricket pitches. *Journal of Geotechnical Engineering, 6*(6), 100–256.

Influence of Flood on the Behavior of Friction Piles

R. S. Athira, S. H. Jasna, K. A. Renjini, Manjima Jayan, Shruthi Johnson, and J. Jayamohan

Abstract Structures are commonly supported on pile foundations at locations having marginal soil. Friction piles are provided wherever the depth to a hard stratum is large. The load settlement behavior of a friction pile entirely depends on the interaction between soil and the pile material at their interface. Recently, our state experienced unprecedented flood which caused wide spread damage to various type of structures and their foundations. It was observed that the majority of failure of structures occurred due to the failure of foundation. This paper investigates the influence of drawdown of water on the load-settlement behavior and stability of friction piles. The results of a series of laboratory-scale load tests to determine the influence of drawdown on the settlement of pile are presented. Laboratory-scale load tests on model pile are carried out in a masonry tank, which has arrangements for pumping in water and drawdown. The influence of rate of drawdown on settlement is studied by varying the discharge of inflow of water. It is observed that the settlement of pile considerably increases due to sudden drawdown.

Keywords Pile foundation · Drawdown · Settlement

1 Introduction

Our state of Kerala experienced unprecedented flood quite recently which caused widespread destructions to various structures. One among the main reasons for the collapse of structures was the failure of foundation due to the sudden movement of water. Damages to the building during inundations can be the result of not only the direct activity of the flood wave and surface water, but also changes in groundwater flow conditions, including the increase of their piezometric level. In the past few decades, many researches have been carried out to investigate the analysis of settlement in foundation due to rise of groundwater. It has been proved that there is an increase in settlement resulting in decrease in bearing capacity during inundation.

R. S. Athira (✉) · S. H. Jasna · K. A. Renjini · M. Jayan · S. Johnson · J. Jayamohan
LBS Institute of Technology for Women, Poojapura, Thiruvananthapuram, India
e-mail: athira04cdlm@gmail.com

© The Editor(s) (if applicable) and The Author(s), under exclusive license to Springer Nature Singapore Pte Ltd. 2021
J. Thomas et al. (eds.), *Current Trends in Civil Engineering*, Lecture Notes in Civil Engineering 104, https://doi.org/10.1007/978-981-15-8151-9_6

Rise of groundwater level is believed to increase the settlement significantly and had been a topic of research for many years. Some of the studies which have been conducted in this field are the effect of Submergence on Settlement and Bearing Capacity by Kazi et al. [1]. They conducted simple laboratory experiments and have shown that the sand bed settles significantly when it is submerged under water for lower values of relative density. Terzaghi (1943) postulated that the submergence of the sand reduces the soil stiffness by half, which in turn doubles the settlement. Stress and pore water pressure changes in partially saturated soils under strip footings had been studied by Mohammed Yousif Fattah et al. (2014). They reported that there are two phenomena governing the behavior of footing represented by settlement (negative vertical displacement) and heave (positive vertical displacement). An increase of load on the foundation will increase the settlement, and the failure surface will gradually extend outward from the foundation in heave behavior.

This paper investigates the impact of floods on the settlement behavior of pile foundation by carrying out a series of laboratory scale load tests. The influences of drawdown on the settlement of friction piles are investigated.

2 Laboratory-Scale Load Tests

The experimental investigation was carried out at the Geotechnical Engineering Research Laboratory of LBS Institute of Technology for Women, Thiruvananthapuram. The details of materials used, experimental setup and methodology are presented below.

2.1 Materials Used

The sand used for granular bed was well-graded sand (SW). Properties of the sand are presented in Table 1.

Table 1 Properties of sand

Property	Sand
Dry unit weight (kN/m^3)	17
Specific gravity	2.3
Cohesion (kPa)	2
Angle of shearing resistance (°)	30

2.2 Experimental Setup

Laboratory-scale load tests are conducted in a test bed and loading frame assembly. The test beds are prepared in a tank which is designed keeping in mind the size of the model pile to be tested and the zone of influence. The dimensions of the test tank are 1000 mm length × 750 mm width × 750 mm depth. Model pile having diameter 12.0 mm and length 60 cm is fabricated with mild steel. The loading tests are carried out in the loading frame fabricated with ISMB 300. The load is applied using a hand-operated mechanical jack of capacity 50 kN. The applied load is measured using a proving ring of capacity 10 kN. Plumbing arrangements were given as two inlet pipe at the bottom of the tank for raising the groundwater level at various discharges at longitudinal and lateral direction, another inlet at top of the tank for giving submergence due to surface flow, and an outlet pipe for measuring the drawdown.

The settlement of model pile is measured using two dial gauges kept diametrically opposite to each other. The model pile is placed exactly beneath the center of loading jack to avoid eccentric loading (Figs. 1 and 2).

Fig. 1 Loading frame

Fig. 2 Plumbing arrangement

2.3 Preparation of Test Bed and Flooding Condition

The initial test is conducted with sand alone in the test tank. Sand is filled in layers of 5 cm thickness and is compacted using a plate vibrator. Piezometers are placed between these layers on opposite side of footing for measuring the head of water. After the preparation of sand bed, the pile is placed at the center of the tank. Two dial gauges are fixed diametrically opposite to each other to measure the deformation. The given load measured by means of a proving ring, simultaneously the settlement measured by means of two dial gauges placed diametrically opposite to each other.

The inlet pipe at the top of the tank is opened and filled the tank 5 cm above the sand bed, thereby creating the surface flooding. Then, the outlet valve is opened for simulating the drawdown condition. The settlement values are then measured from the dial gauges corresponding to the time intervals. The corresponding head can be measured by means of the level of water in the piezometers (Fig. 3; Table 2).

3 Results

See Figs. 4, 5 and 6.

Fig. 3 Test setup for laboratory-scale load test

Table 2 Discharge parameters varied in laboratory-scale load test

Variation of discharge	Full	½
Lateral discharge (10^{-6} m^3/s)	1.63	1.28

4 Discussion

In case of sudden drawdown, the discharge on the lateral direction is varied and it is observed that settlement increases with increase in time. The increase in settlement is due to hydrostatic pressure that develops with increase in head.

5 Conclusion

In drawdown condition, settlement increases with time. Settlement increases at a rapid rate for full discharge condition.

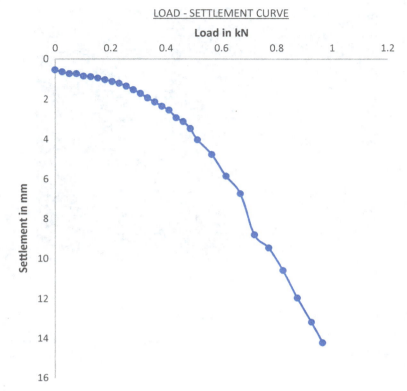

Fig. 4 Load-settlement curve

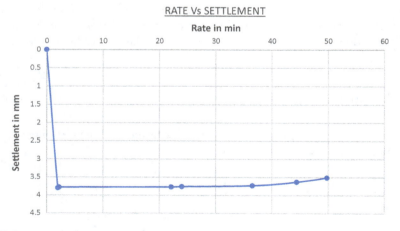

Fig. 5 Rate versus settlement (full discharge)

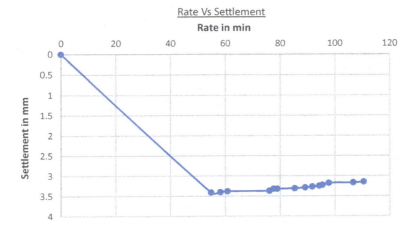

Fig. 6 Rate versus settlement (half discharge)

References

1. Kazi, M., Shukla, S. K., & Habibi D. (2015). Effect of submergence on settlement and bearing capacity of surface strip footing on geotextile-reinforced sand bed. *International Journal of Geosynthetics And Ground Engineering, 1.* Article number: 4.
2. Fattah, M. Y., Ahamed, M. D., Mohammed, H. A.. (2014). Stress and pore water pressure changes in partially saturated soils under strip footings. *International Conference for Engineering Sciences*, University of Mustansiriya, Bagdhad.
3. Terzaghi K. (1943). Theoretical Soil Mechanics, John Wiley and Sons. New York,NY,USA,1943.

Feasibility Study of Using Coir Geotextiles in Permeable Pavement Construction for Stormwater Management

Mohan Kavitha, Subha Vishnudas, and K. U. Abdu Rahiman

Abstract Stormwater management is a challenging task for countries all around the world due to rapid urbanization and climate change. Different low impact development (LID) practices have been adopted around the world to tackle the situation. Permeable pavements are a widely accepted solution to stormwater management and groundwater recharge. The structure of permeable pavements incorporates a layer of geotextile to stabilize the reservoir course. However, the other properties of the geotextile layer that includes drainage and filtration also need to be looked into. This paper tries to review different properties of geotextiles when incorporated into the soil and check the feasibility of using coir geotextiles in the permeable pavement structure.

Keywords Stormwater management · Permeable pavements · Coir geotextiles · LID practices

1 Introduction

Rapid urbanization has brought along its wake a lot of problems, especially from the point of view of drainage. As the built-up areas increased, the natural drainage characteristics of the lands were affected, leading to flooding and unavailability of water fit for human conception. Scarcity of water has highlighted the importance of rainwater harvesting and groundwater recharge. Low impact development (LID) practices and sustainable urban drainage systems (SUDS) are now becoming popular

---

M. Kavitha (✉)
Department of Civil Engineering, Rajagiri School of Engineering and Technology, Kochi, India
e-mail: kavitham@rajagiritech.edu.in

S. Vishnudas · K. U. Abdu Rahiman
Division of Civil Engineering, Cochin University of Science and Technology, Kochi, India
e-mail: v.subha@cusat.ac.in

K. U. Abdu Rahiman
e-mail: arku@cusat.ac.in

around the world for being able to counter the ill effects of urbanization like pollution and drainage problems without compromising on the development of the region.

Permeable pavements are a long existing LID practice that helps reduce surface runoff and augments groundwater recharge. The pavements, incorporated in low traffic areas and parking lots, act as a storage reservoir, storing the excess rainwater in its reservoir course and thus preventing the water from flowing out as surface runoff.

Many studies have been conducted to quantify the effectiveness of permeable pavements for small as well as large storms. Experimental studies carried out by [1] showed that permeable pavements reduced runoff by 93% compared to an asphalt pavement subjected to the same rainfall even after being constructed above clayey soils. A performance comparison of four different types of permeable pavements was carried out by [2], and though their response was significantly different for smaller storms, their responses were comparable during larger storms and much better than asphalt pavements.

Efficiency of permeable pavements for draining rainwater largely depends on the storage capacity of the reservoir course. Clogging of the pavements by fine sediments that entered via the gaps in the pavement was found to be the biggest source responsible for reduction in their lifespan. It was pointed out by [3] that permeable pavements were not given BMP credits due to the fact that it is prone to clogging. It was also examined by [4] that the life of these pavements cannot be predicted accurately, as the sources of clogging vary for different sites. Reference [5] discussed about the development of a 'clogging front' that forms as runoff passes over already clogged regions and infiltrates downstream, clogging it with more sediment and organic matter as it moves forward.

1.1 Structure of a Permeable Pavement

The structure of a permeable pavement illustrated by [6] is shown in Fig. 1. The permeable paving surface is underlain by a porous bedding media and separated from the lower reservoir base course by a geotextile separation layer. The reservoir base course is separated from the existing soil subgrade by another geotextile layer.

Each of the components of the permeable pavement has functions of their own. After water travels through the gaps in the porous paving, it passes through the bedding course, usually comprising of sand or gravel. The upper geotextile filters the water of impurities, and the water passes on into the reservoir sub-base course, where it is stored till it slowly infiltrates into the existing soil subgrade and eventually recharges the groundwater. The lower geotextile layer is provided to separate the base course from the existing subgrade and stabilize the pavement structure. A perforated underdrain can be provided to collect the water stored in the reservoir course, in case it is not desired by the designers for the water to enter the subgrade and weaken its structure.

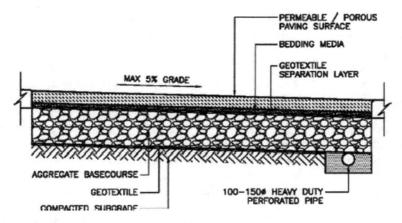

Fig. 1 Longitudinal section of a permeable pavement

The functions of different components of the permeable pavements have been studied by many authors. The most interesting part though is how many of them seem to overlook the functions of the geotextile layers. Many experimental studies on permeable pavements have been carried out in the absence of a geotextile layer, deeming them as optional. This paper tries to review different studies that have been conducted on the geotextile layer of the permeable pavement structure. An attempt is made to check whether the synthetic geotextile layer can be replaced by naturally available geotextiles like coir by reviewing the essential characteristics of both the fibers based on the literatures available.

2 The Geotextile Layer in Permeable Pavements

2.1 The Function of the Geotextile Layer—Review of Studies Conducted

The use of geotextiles in permeable pavements is fivefold—drainage, filtration, reinforcement, separation and inter-layer functions.

It was reported by [7] that permeable pavements act as in-situ aerobic bio-reactors to break down hydrocarbons present in pollutants like engine oils and improve the quality of infiltrating water. Some researchers such as [8] believe that the geotextile layer plays an important role in improving the quality of water that passes through the permeable pavement structure. It was considered to be effective in trapping pollutants and biodegrading it within the system.

Many researchers have studied about the biodegradation of pollutants at the site of the upper geotextile of the permeable pavement system. They discovered that the one of the major source of pollutants into the water that moves through the permeable pavement was engine oil and similar hydrocarbons. The biodegradation of the oil particles could be carried out by the microbial activity on the upper geotextile. However, the unavailability of nutrients like N, K and P in this system affects the sustained health of the bacteria required for oil biodegradation. In order to overcome this situation, [9] examined the use of a self-fertilizing geotextile—where treated phosphate beads were incorporated into the geotextile leading to slow release phosphate into the system, which will help the bacteria survive.

One of the most significant studies carried out on the importance of geotextile layer to the structure of the permeable pavement was the Abertay study ([10, 11]). The researchers did an experiment in elevated test rigs under accelerated time scale to check the effectiveness of pollutant removal capacity of geotextiles. However, [12] questioned the validity of the study due to their decision to accelerate the timeline of 10 years' worth pollutant loads to 2 months, arguing that it did not give enough time for the natural systems to have settled in place. A valid statistical analysis could not be carried out during the Abertay study, due to the variability in data, and hence, a clear correlation of whether a geotextile layer is responsible for the removal of pollutants could not be reached.

It was pointed out by [12] that geotextile layer, while allowing water to pass through, retained the sediment on it, which will end up decreasing the heavy metal pollution of the water, since most of the heavy metals are carried into the water by the sediments on the paving surfaces.

The review of the studies conducted on the geotextile layer incorporated in a permeable pavement, leaves a lot of questions unanswered. The number of attempts made to correlate the quantity and quality of the stormwater runoff with the presence of a geotextile layer in a permeable pavement is very less and leaves a lot of scope for further experimentation.

3 The Characteristics of Geotextiles in Pavement Construction

While the studies relating to the performance of geotextile layer in permeable pavements are very few, many studies are available on the characteristics of geotextile applied to pavement construction—especially with regards to its reinforcement and strength characteristics.

Some index properties are available that can help determine the characteristics of geotextiles. These include mass per unit area, apparent opening size, falling head permittivity, puncture resistance, trapezoidal tear, grab tensile and wide-width tensile strengths.

Mass per unit area is determined for the purpose of quality control. It determines the specimen conformance.

Apparent opening size signifies the ability of a geotextile to retain fines. According to experiments carried out by [13], if the apparent opening size (O_{95}) of the geotextile is less than 85% size (d_{85}) of the retained soil, it would perform satisfactorily. For fine soils in suspension, it was concluded that the ratio of O_{95} to d_{85} should be limited below 0.5.

Significance of puncture resistance lies in the fact that geotextiles exhibiting greater puncture resistance would theoretically provide greater resistance to the punching action of aggregates above the subgrade.

Trapezoidal tear test tries to measure the tensile force at which a tear in the geotextile will start to propagate and cause damage to the geotextile as a whole. Wide-width tensile test helps determine the tensile strength of the specimen when gripped across the whole width of the specimen, while the grab tensile strength helps determine the tensile strength of the geotextile, when gripped only for a part of the entire width. Wide-width tests are carried out in both machine and cross-machine directions, in order to determine the best way to orient the geotextile. Elongation of the geotextile is also recorded while conducting these tests. It is a good sign, if the geotextile elongates more, showing that it can resist the tears more.

Common index properties of 17 geotextiles (which included different varieties of woven and non-woven geotextiles) were conducted by [14]. The woven geotextiles tested included slit-film geotextiles and combination geotextiles of monofilament fibrillated, monofilament and slit-film weaves. The non-woven geotextiles tested included needle punched as well as heat bonded geotextiles. The mean values of the obtained results are given in Table 1.

In addition to the above properties, the apparent elongation of woven geotextiles at failure was found to be 26%, while that of non-woven geotextiles was found to be 82%. The results of the wide-width tensile tests conducted by [14] are given in Table 2.

From the tables, it is clear that though the tensile strengths of woven geotextiles are much greater than non-woven geotextiles, the permittivity of non-woven geotextiles is greater than their woven counterparts. When selecting an appropriate geotextile to be used in a permeable pavement, an engineer should be able to weigh in the pros and cons of both types of geotextiles to come to an answer. While it would be preferred to go for a non-woven geotextile over structurally sound subgrades, woven geotextiles would be preferred over weaker soils such as silt or clay. Among woven geotextiles, it would be better to avoid slit-film geotextiles, due to their low permeability values, which can adversely affect the recharge of groundwater.

Table 1 Mean values of the common index properties of geotextiles for use in pavement construction

Type of geotextile		Mass per unit area (g/m²)	AOS (mm)	Permittivity (s⁻¹)	Puncture resistance (kg)	Trapezoidal tear test—maximum tensile strength (kg)	Grab tensile test—breaking strength (kg)
Woven	Slit-film	221	0.270	0.10	53.07	62.14	158.76
	Combination	334	0.518	0.39	67.13	95.70	215.91
Non-woven	Needle punched	227	0.124	1.21	48.53	43.54	90.26
	Heat bonded	153	0.166	1.03	26.76	35.38	87.54

Table 2 Mean values of wide-width tensile test results of woven and non-woven geotextiles

Type of geotextile		Tensile strength (kg)		Elongation (%)	
		MD	CMD	MD	CMD
Woven	Slit-film	93.44	108.86	20	12
	Combination	156.04	156.49	16	15
Non-woven	Needle punched	35.83	47.17	53	44
	Heat bonded	24.49	25.40	63	58

[a]MD—machine direction, CMD—cross-machine direction

4 Characteristics of Coir Geotextiles

Coir geotextile is a natural alternative to the geosynthetic materials that are being used in pavement construction. Coir fiber has been successfully tested for the engineering properties of separation, filtration, reinforcement and drainage by many researchers.

Studies have been conducted to establish the reinforcement and soil stabilization characteristics of coir by [15–17]. The characteristics of coir have been found to be comparable to the performance of geosynthetic materials as far as reinforcement is concerned. Coir has been found to have the highest tensile strength among other natural fibers like cotton, jute and sisal. Its ability to retain strength after installation is much higher than the other natural alternatives. Coir geotextiles are today widely used for strengthening embankments, roads and even the soil below the foundations.

Various studies have been conducted to study the durability of coir geotextile used for reinforcement purposes. In an experimental study conducted by [18], where coir geotextile was used to stabilize the banks of a village pond, it was examined that coir retained 19% of the strength of a fresh sample after 9 months. Another important point that discussed was the reduction in nutrient loss observed in the soil that has been treated with coir geotextile and compared to soil that was not. This had been attributed to the nutrients being made available to the soil on biodegradation of the coir itself.

The ability of coir geotextile to retain moisture content was established by [19]. The coir geotextile was used in the study as an alternative to providing bench terraces to stabilize slopes for cultivation. It was found that the coir not only stabilized the slopes, but was also the source of high moisture retention, which increased the moisture availability in the soil. The study also observed that the higher percentage of volumetric moisture content in the plots treated with coir geotextiles may be due to the mesh opening in them which provides a large number of porous check dams in the soil, leading to the sediment settling and water passing through the matting. It was claimed by [20] that the longevity of geotextiles depended on their material composition and natural geotextiles last for 2–4 years. Its high strength can be attributed to its high lignin content which also makes it resilient and durable.

Coir fiber modified with hydrogen peroxide was found to act as an adsorbent of heavy metal ions like Ni, Zn and Fe from their aqueous solutions by [21].

The strength characteristics of coir geotextiles were discussed by many researchers. According to [22], the maximum values of tensile strength of different varieties of coir tested ranged up to 30 kN/m and tensile strain at failure ranged upto 40%. These characteristics are comparable to available high strength geosynthetic products in the market.

4.1 Possibilities of Using Coir in Permeable Pavements

There have not been any studies to describe the pros and cons of using natural geotextiles in the context of permeable pavements. Strength characteristics of coir have been widely studied, leading to the use of these abundantly available natural geotextiles as reinforcement in paved and unpaved roads. However, natural geotextiles like coir when used for reinforcement are intended only till the time a vegetative cover can replace it and hold the soil together.

Unlike unpaved roads or unstable slopes, the property of strength is not that important in the case of a geotextile layer inserted in a permeable pavement, as it is intended mainly for the purpose of separation.

The ability of coir to showcase higher strength compared to other geotextiles in the wet condition and its ability to degrade into nutrients in the soil will be able to augment the quality and quantity of groundwater recharge. The filtration characteristics of coir have also been reported to be of engineering significance by many researchers. The possibility of modified coir being able to adsorb heavy metals also cannot be overlooked.

The durability of coir is one of the problems pointed out by the researchers. However, the effect of the fiber when subjected to an environment like a permeable pavement that gets affected by temperature and pH not similar to soil will require extensive experimental study. Hence, coir which has exhibited certain promising characteristics should definitely be studied as a sustainable alternative to other geosynthetic materials to be used as part of the permeable pavement structure.

5 Conclusions

Permeable pavements have been an acceptable solution for groundwater recharge across the globe. Even though the amount of recharge is found to be largely affected by clogging as the time progresses, it is still found to be effective. In addition to quantity, the quality of the infiltrated water too can be improved by the use of geotextile layer in the pavements. However, except for its tensile strength, a lot of studies have not been conducted to examine the performance of the geotextile layer of the permeable pavement. Even, geotextile functions like filtration and drainage have not been examined by many researchers.

Coir geotextiles, a product obtained from coconut husk, that are abundantly available in tropical countries like India, have been extensively studied and used as reinforcement in pavements and unstable slopes. To extend their use into the structure of permeable pavements, their properties of filtration, pollutant removal and drainage should be studied in detail by means of extensive experimentation. If it is found to be successful, then it would boost the possibilities of groundwater recharge and also promote the use of sustainable materials.

References

1. Dreelin, E. A., Fowler, L., & Carroll, C. R. (2006). A test of porous pavement effectiveness on clay soils during natural storm events. *Water Research, 40*(4), 799–805.
2. Collins, K. A., Hunt, W. F., & Hathaway, J. M. (2008). Hydrologic comparison of four types of permeable pavement and standard asphalt in eastern North Carolina. *Journal of Hydrologic Engineering, 13*(12), 1146–1157.
3. Bean, E. Z., Hunt, W. F., & Bidelspach, D. A. (2007). Field survey of permeable pavement surface infiltration rates. *Journal of Irrigation and Drainage Engineering, 133*(3), 249–255.
4. Pratt, C. J., Mantle, J. D. G., & Schofield, P. A. (1995). UK research into the performance of permeable pavement, reservoir structures in controlling stormwater discharge quantity and quality. *Water Science and Technology, 32*(1), 63–69.
5. Pezzaniti, D., Beecham, S., & Kandasamy, J. (2009). Influence of clogging on the effective life of permeable pavements. *Water Management, 162*(WM3), 211–220.
6. Fassman, E. A., & Blackbourn, S. (2010). Urban runoff mitigation by a permeable pavement system over impermeable soils. *Journal of Hydrologic Engineering, 15*(6), 475–485.
7. Pratt, C. J., Newman, A. P., & Bond, P. C. (1999). Mineral oil biodegradation within a permeable pavement: Long term observations. *Water Science and Technology, 39*(2), 103–109.
8. Nnadi, O. E., Newman, P. A., & Coupe, J. S. (2014). Geotextile incorporated permeable pavement system as potential source of irrigation water : Effects of re-used water on soil, plant growth and development. *CLEAN—Soil Air Water, 42*(2), 125–132.
9. Spicer, G. E., Lynch, D. E., Newman, P. A., & Coupe, J. S. (2006). The development of geotextile incorporating slow release phosphate beads for the maintenance of oil degrading bacteria in permeable pavements. *Water Science and Technology, 54*(6–7), 273–280.
10. Mullaney, J., Jefferies, C., & Mackinnon, E. (2011). The performance of block paving with and without geotextile in the sub-base. In *Proceedings of SUDSnet National and International Conference in Dundee, UK,* May 11–12, 2011
11. Mullaney, J., Rikalainen, P., & Jefferies, C. (2012). Pollution profiling and particle size distribution within permeable paving units—With and without geotextile. *Management of Environmental Quality, 23*(2), 150–162.
12. Scholz, M. (2013). Water quality improvement performance of geotextiles within permeable pavement systems: A critical review. *Water, 5,* 462–479.
13. Narejo, D. B. (2003). Opening size recommendations for separation geotextiles used in pavements. *Geotextiles and Geomembranes, 21*(4), 257–264.
14. Jersey, S. R., & Tingle, J. S. (2007). Geotextile response to common index tests. *Transportation Research Record, 2*(1989), 102–112.
15. Lekha, K. R. (2004). Field instrumentation and monitoring of soil erosion in coir geotextile stabilised slopes. *Geotextiles and Geomembranes, 22*(5), 399–413.
16. Lekha, K. R., & Kavitha, V. (2006). Coir geotextile reinforced clay dykes for drainage of low-lying areas. *Geotextiles and Geomembranes, 24*(1), 38–51.
17. Subaida, E. A., Chandrakaran, S., & Sankar, N. (2008). Experimental investigations on tensile and pullout behaviour of woven coir geotextiles. *Geotextiles and Geomembranes, 26*(5), 384–392.

18. Vishnudas, S., Savenije, H. H. G., Van der Zaag, P., Anil, K. R., & Balan, K. (2005). Experimental study in coir geotextiles in watershed management. *Hydrology and Earth System Sciences Discussions, 2*(6), 2327–2348
19. Vishnudas, S., Savenije, H. H. G., Van der Zaag, P., & Anil, K. R. (2012). Coir geotextile for slope stabilization and cultivation—A case study in a highland region of Kerala, South India. *Physics and Chemistry of the Earth, 47–48*, 135–138
20. Rickson, R. J. (2006). Controlling sediment at source: An evaluation of erosion control geotextiles. *Earth Surface Processes and Landforms, 31*(5), 550–560.
21. Shukla, S. R., Pai, R. S., & Shendarkar, A. D. (2006). Adsorption of Ni(II), Zn(II) and Fe(II) on modified coir fibres. *Separation and Purification Technology, 47*(3), 141–147.
22. Rao, V. G., & Dutta, R. K. (2005). Characterisation of tensile strength behaviour of coir products. *Electronic Journal of Geotechnical Engineering, 10*(B).

Assessment of Effect of Filler in the Properties of Cement Grout

A. B. Kavya and S. R. Soorya

Abstract The scarcity of land for construction purposes nowadays demands the use of economical techniques such as grouting for strengthening the available soils especially loose sandy soils. Conventional material adopted for suspension grouting is OPC which is not much cost-effective. So the use of certain additives as partial replacement to cement is gaining more importance. Therefore, the present study focuses on the use of metakaolin as a partial replacement to cement as it is a good pozzolanic material. Here metakaolin is added at about 10% by dry weight of cement. The properties of the grout such as viscosity and bleeding are studied for different water–binder ratios of 9:1, 8:2, and 7:3. Also, one of the most important properties of the grout to give maximum grouted volume with minimal weight without disturbing the soil skeleton is evaluated. The results indicated that the addition of metakaolin improved the viscosity and reduced the bleeding of the grout. The best result was obtained for water–binder ratio 8:2. The grouted zone volume also showed an increase with 10% metakaolin addition. Thus, the effectiveness of grout was found to be more at water–binder ratio of 8:2. Based on these results, metakaolin modified cement grout can be studied in future for its ability to impart strength and improve the permeability characteristics of sandy soils.

Keywords Grouting · Metakaolin · Viscosity · Portland pozzolanic cement

1 Introduction

The constructional activities in the coastal belt of our country often demand deep foundations because of the poor engineering properties and the related problem arising from weak soil at shallow depths. The very low shearing resistance of the

---

A. B. Kavya (✉) · S. R. Soorya
Department of Civil Engineering, Marian Engineering College, Trivandrum, India
e-mail: kavyanair071@gmail.com

S. R. Soorya
e-mail: sooryasree179@gmail.com

© The Editor(s) (if applicable) and The Author(s), under exclusive license to Springer Nature Singapore Pte Ltd. 2021
J. Thomas et al. (eds.), *Current Trends in Civil Engineering*,
Lecture Notes in Civil Engineering 104, https://doi.org/10.1007/978-981-15-8151-9_8

foundation bed causes local as well as punching shear failure. Hence, the structures built on these soils may suffer from excessive settlements. Strengthening of these loose sandy soils at shallow depth through economical techniques such as grouting is a possible solution. Grouting often has to serve the primary purpose of filling the voids or replacing the existing fluids in voids with a view to improve engineering properties of the grouted medium.

Commonly adopted grouting material is cement as it offers improved properties for a sand medium. The use of various additives along with cement proves to improve the efficiency of cement grout. Hwang et al. [1] investigated the applicability of bentonite suspensions treated with sodium pyrophosphate as permeation grout by measuring the yield stress and viscosity. Dayakar et al. [2] studied the effect of permeation grouting using cement grout in sandy soil with different water–cement ratios. Improvement in bearing capacity of the sample was studied by conducting plate load test after 3 and 7 days curing. For best result for the grouting process, the selection of grout must be more suitable to the problem in the terms of viscosity, setting time, strength, but the method of how to distribute grout in the soil is also an important factor and it is necessary to make a correct choice of equipment of grouting, distance between grout holes, length of injection passes, a number of grouting phases, grouting pressure, and pumping rate.

Ozgurel and Vipulanandan [3] conducted experiments on different gradations of sand which were grouted using acrylamide chemical grout in order to investigate the effect of grain size distribution of sand on the mechanical properties and permeability of grouted sand. He observed that groutability was greatly influenced by the fines content. Sometimes grouts with Portland cement need adding small amounts of additives, which are used for specific purposes such as increasing fluidity, retarding sedimentation, and controlling the set time. Penetrability of suspended solids grouts will be limited by the ratio of opening size to grain size, which must be three or more for successful grouting process, [4]. Fillers materials with cement grout are used primarily for economic reasons as a replacement material where substantial quantities of grout are required to fill large cavities in rock or in soil (trenches, cavities, to stem boreholes, shafts, and tunnels). Almost any solid substance that is pumpable is suitable as the filler in grout to be used in non-permanent work. For permanent work, cement replacements should be restricted to mineral fillers. Water–binder ratio is one of the important parameters of a grout mix. When the value of water–binder ratio is very low, it affects the properties of grout such as reducing bleeding, decrease in workability, improving strength, and durability [5].

Table 1 Properties of sand

Properties	Value
Specific gravity	2.64
Effective size D_{10}, mm	0.3
Uniformity coefficient, Cu	1.53
Coefficient of curvature, Cc	0.992
Coefficient of permeability, k (m/s)	4.25×10^{-5}
Void ratio, e	0.504
Bulk density, (kg/m^3)	1627
Porosity, n	0.335
Angle of internal friction, ϕ (deg)	39.11
Cohesion (kPa)	0.2
Classification (IS)	SP

2 Materials

2.1 Sand

The sand used in this study is collected locally from Neyyatinkara, Thiruvananthapuram District. The properties of the soil are shown in Table 1.

2.2 Cement

Portland pozzolanic cement of grade 33 conforming to IS 1498 is used for preparing grout. It was collected from a local supplier. Table 2 shows the properties of cement used.

Table 2 Properties of cement

Properties	Value
Grade	33
Fineness (%)	6.5
Standard consistency (%)	28
Initial setting time (min)	85
Final setting time (min)	420

2.3 Metakaolin

Metakaolin used for the study was collected from English India Clay Ltd., Veli. It typically contains SiO_2 and Al_2O_3, but other oxides present in small amounts include Fe_2O_3, TiO_2, CaO, and MgO. Metakaolin particles are generally one-half to five microns in diameter, an order of magnitude smaller than cement grains. Due to the controlled nature of the processing, metakaolin powders are very consistent in appearance and performance. Table 3 shows the properties of metakaolin used.

3 Methodology

The grout was prepared using cement with and without metakaolin additive. The metakaolin was added at 10% by dry weight of cement. The study was carried out in three different water–binder ratios of 9:1, 8:2, and 7:3.

A Marsh funnel test conforming to IS 14343:1996 was conducted to determine viscosity. The result to the nearest second is the Marsh funnel viscosity. For grout slurries, the Marsh funnel viscosity has a good relationship with viscosity in centipoise determined by a rotational viscometer. The relationship is nearly straight line in the range of 30–40 s. The test setup is shown in Fig. 1.

Bleeding test was carried out for the prepared grout as per ASTM-C940. The readings were recorded at 15 min interval for first 60 min, and thereafter at hourly intervals, until no further bleeding is observed. According to ASTM-C937, the test should be discontinued after 3 h. At last, the bleed water is decanted and volume is obtained. The bleeding of grout mix was obtained using the expression, where V_1 is the volume of sample at the beginning of test (ml), V_2 is the volume of sample at prescribed intervals (ml), and V_g is the volume of grout portion of sample at prescribed intervals (ml). Figure 2 shows the bleeding test setup.

Table 3 Properties of Metakaolin

Properties	Value
Specific gravity	2.6
Liquid limit (%)	85
Plastic limit (%)	33
Plasticity index	52
Shrinkage limit (%)	21
Clay (%)	80
Silt (%)	12
Sand (%)	8
Classification	CH

Fig. 1 Viscosity test setup

Fig. 2 Bleeding test setup

To determine ability of grout to give maximum volume with minimal weight without disturbing the soil skeleton and to check how these materials can fill the soil voids, cylinder model test is carried out. A PVC cylinder of height 0.3 m and diameter 0.15 m are used for the test. The cylinder was filled to a depth of 0.25 m with loose sand. The slurry was injected to depths 0.125 and 0.17 m for each water–cement ratio with and without filler materials. The injection of slurry was carried out with

Fig. 3 Cylinder model test setup

a grouting set up without disturbing the soil skeleton. After one day of curing time, the approximate dimensions of intact grouted mass were determined using water displacement method. Grouting setup shown in Fig. 3 consists of a grout chamber with air compressor, grouting nozzle, and a regulating valve.

4 Results and Discussions

4.1 Viscosity

The marsh funnel test conducted on the grout samples prepared indicates that the addition of 10% metakaolin to cement showed a considerable value of viscosity suitable for grouting a sand medium. Also with an increase in the water–binder ratio, the viscosity value is found to decrease. Thus, the addition of 10% metakaolin is found to produce more effective grout mix compared with slurry prepared with cement alone. This is due to the increased fineness and plasticity characteristics of the metakaolin particle compared to cement particles. Viscosity test results are shown in Fig. 4.

4.2 Bleeding

The bleeding is found to be less for cement grout sample modified with 10% metakaolin. Also with a decrease in water–binder ratio, bleeding is found to decrease. This is due to the increased cement surface offered by the metakaolin added to the cement. Moreover thinner grouts settle more rapidly, the addition of metakaolin to

Assessment of Effect of Filler in the Properties of Cement Grout

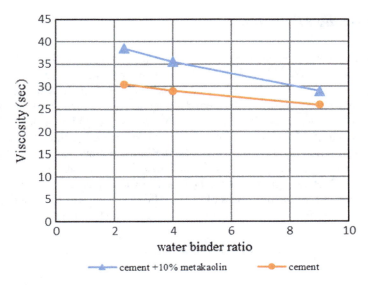

Fig. 4 Variation of average viscosity with water–binder ratio

cement enhances the settlement and thereby reduces bleed. Figure 5 shows the test result.

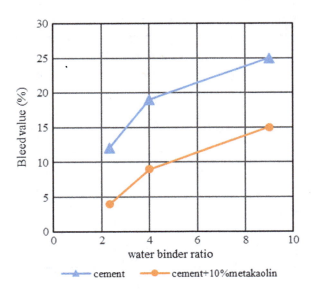

Fig. 5 Variation of bleed value with water–binder ratio

4.3 Cylinder Model Test

The test conducted indicated the effectiveness of 10% metakaolin added cement grout in grouting more volume of sand. For cement grout without additive, at lower water–binder ratio the volume of grouted zone obtained is more with the addition of 10% metakaolin to cement grout found to increase the volume of grouted zone at higher water–binder ratios. The variation with depth is also studied which suggested that at a depth of 17 cm volume obtained is less compared to that at 12.5 cm for water–binder ratio 7:3. For higher water–binder ratios, an inverse trend is observed. Thus, grout slurry prepared with cement and 10% metakaolin at water–binder ratio 8:2 is much effective in grouting a sand medium at varying depths. This is due to the increased surface area and fineness of metakaolin which enables the grout in filling more voids. Figures 6, 7, and 8 illustrate the cylinder model test results.

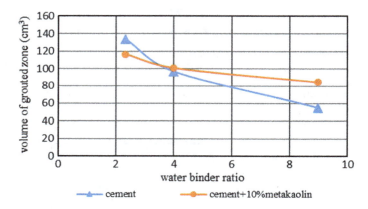

Fig. 6 Variation of volume of grouted zone for depth = 0.125 m

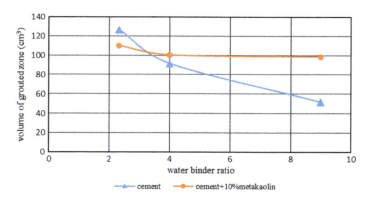

Fig. 7 Variation of volume of grouted zone for depth = 0.17 m

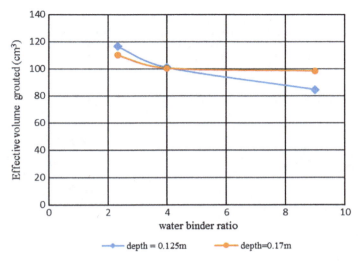

Fig. 8 Variation of grouted zone volume with depth

5 Conclusions

- The test results showed that the addition of 10% metakaolin as a partial replacement to cement will result in a more effective grout slurry.
- The viscosity values for metakaolin modified grout were obtained in a range of 30–40 s which suggests it as a good grout. With an increase in the water–binder ratio, the viscosity value is found to decrease.
- Bleeding found to decrease with addition of metakaolin which is a reliable result.
- More grouted zone volume was obtained for metakaolin modified grout than for plain cement grout at the same water–binder ratios.
- At higher water–binder ratios, more effective grouting is possible for greater depths of 0.17 m.
- The optimum water–binder ratio was observed as 8:2 for 10% metakaolin modified cement grouts.

References

1. Hwang, H., Yoon, J., Rugg, D., & El Mohtar, C. S. (2010). Hydraulic conductivity of bentonite grouted sand. *Journal of Geotechnical and Geoenvironmental Engineering, ASCE, 13*(9), 1–10.
2. Dayakar, P., Venkat, K. R., & Raju, K. V. B. (2012). Study on permeation grouting using cement grout in sandy soils. *Journal of Mechanical and Civil Engineering, 4*(4), 5–10.
3. Ozgurel, H. G., & Vipulanandan, C. (2005). Effect of grain size distribution on permeability and mechanical behaviour of acrylamide grouted sand. *Journal of Geotechnical and Geoenvironmental engineering, ASCE, 131*(12), 57–65.

4. Karol, R. H. (1982). Seepage control with chemical grouts. In W. H. Baker (Ed.), *Proceedings of the Conference on Grouting in Geotechnical Engineering*, New Orleans, ASCE, New York.
5. Bruce, D. A. (2002). Anchors, Micropiles, Rock Grouting and Deep Mixing: A Decade of Progress in the United States.

High-Strength Geopolymer Mortar Cured at Ambient Temperature

Job Thomas and N. J. Sabu

Abstract Addition of 12 percentage of OPC along with fly ash acts as the main source of Si and Al for the geopolymerisation. It is seen that the inclusion of OPC helps in geopolymerisation without heating. A mortar strength up to 105.5 MPa was prepared in this study. A total of 81 mixed geopolymer mortar mixes was selected. The molarity of NaOH, ratio of sodium silicate (Na_2SiO_3) to sodium hydroxide (NaOH), alkali–binder ratio and cement content were varied. The 7 and 28 days cube compressive strength were determined. The strength prediction models developed in this study are found to be in good agreement with the experimental data.

Keywords Compressive strength · Geopolymer · Ambient curing · Fly ash · OPC

1 Introduction

The increased demand of cement production leads to the excessive emission of carbon dioxide and contributes serious environmental problems. Geopolymer mortar is a potential construction material to be considered in future to minimise the environmental damages. However, the use of conventional geopolymer mortar or concrete can be considered only for the manufacture of precast elements because heat curing is an essential treatment of its production process. Davidovits [1] introduced the geopolymer, the binding material of which essentially consists of silica and alumina in 1978. Geopolymerisation is taking place due to the activation of aluminosilicate by alkali activators such as sodium hydroxide (NaOH) or potassium hydroxide (KOH) and sodium silicate (Na_2SiO_3) or potassium silicate (K_2SiO_3). In construction industry, the geopolymer binders can replace the traditional binders and can be used for the production of structural elements [7].

J. Thomas
Department of Civil Engineering, School of Engineering, Cochin University of Science and Technology, Cochin, Kerala 682022, India

N. J. Sabu (✉)
Research Scholar, Department of Civil Engineering, CUSAT, Cochin, Kerala 682022, India
e-mail: sabunj@gmail.com

© The Editor(s) (if applicable) and The Author(s), under exclusive license to Springer Nature Singapore Pte Ltd. 2021
J. Thomas et al. (eds.), *Current Trends in Civil Engineering*,
Lecture Notes in Civil Engineering 104, https://doi.org/10.1007/978-981-15-8151-9_9

The use of geopolymer concrete in in-situ construction is limited because of the difficulty in heat curing at sites. To mitigate this difficulty, addition of little quantity of high reactive pozzolanic binder is suggested and practised [9]. Somna et al. [14] studied the role of using ground fly ash in the alkaline environment for the development of compressive strength of geopolymer. The use of mineral admixtures such as silica fume, rice husk ash, metakaolin and blast furnace slag to avoid heat curing of geopolymer was reported in the literatures [13]. In this study, small quantity of ordinary Portland cement (OPC) is used to improve the properties of geopolymer concrete. The aluminosilicate geopolymer gel formed along with the calcium silicate hydrate (CSH) gel provides strength of the matrix [2, 3, 16]. Rashad [13] pointed out that the geopolymerisation process can be enhanced by incorporating the materials containing calcium oxide. The studies by Nath and Sarkar [10] indicated that addition of little amount of Portland cement helps in the setting and hardening of geopolymer concrete without heat curing. In this study, OPC included geopolymer mortar is prepared and tested to evaluates the potential of using it in on-site construction application.

There are three main stages in the geopolymerisation process, namely dissolution of oxide minerals from the source material under highly alkaline condition, transportation of dissolved oxide minerals followed by coagulation, polycondensation to form three-dimensional network of silicoaluminate structures. Fly ash is the source material from which silica and alumina are utilised for the polymerisation process.

2 Experimental Programme

The materials were tested, and the mix design is carried out.

2.1 *Materials*

Low calcium fly ash, cement, sodium hydroxide, sodium silicate, water and fine aggregate were used. All materials were tested, and properties were determined. Low calcium fly ash (Class F) conforming to IS 3812-1981 [6] and having specific gravity of 1.98 was used. OPC 53-grade cement conforming to IS 12269-1987 [4] having specific gravity of 3.10 was used. In this study, the OPC 53-grade cement is used as the additive to avoid the heat curing and compressive strength of geopolymer concrete. The laboratory grade (97%) flakes of sodium hydroxide and 52-grade sodium silicate were used. Crushed granite fines having specific gravity of 2.62 and conforming zone II of IS 383-1970 [5] were used as fine aggregate.

2.2 Details of Mix Proportion

Variables considered in this study are OPC content, molarity of sodium hydroxide, ratio of sodium silicate-to-sodium hydroxide and alkali–binder ratio. Fly ash and OPC together act as the binder in the geopolymer system. The OPC content 6, 9 and 12% by weight of total binders and the sodium hydroxide of molarity 8, 12 and 16 M were used. The ratio of sodium silicate-to-sodium hydroxide was varied between 2.2 and 2.8, and the liquid alkaline activator to fly ash ratio of 0.40, 0.45 and 0.5 was used.

The mix proportioning of mortar was carried out based on density method proposed by Rangan [12] and Nath and Sarkar 2015 [10]. The target density of fresh geopolymer mortar mix was 2200 kg/m^3. The weight of binder content of 33% by weight was used in this mix design. The designation of sample is represented by cement content/molarity of sodium hydroxide/sodium silicate-to-sodium hydroxide ratio/alkali–binder ratio. For example, the designation 6/8/2.2/0.40 represents a mix with molarity of 8, sodium silicate-to-hydroxide ratio of 2.2 and alkali–binder ratio of 0.40. The designation and weight of the constituent materials are given in Table 1.

2.3 Preparation of Specimens

Initially, NaOH solid flakes were dissolved in water to get the required molarity solution. The 1 M solution is prepared by mixing 40 g of NaOH solids in 1 L of solution. The solution is allowed to cool to room temperature. The sodium hydroxide solution is added to the required quantity of sodium silicate solution and stirred thoroughly. This forms the alkali activator, which is allowed to cool for about 1 h in water bath. The binder materials, namely fly ash and cement, mixed dry. The fine aggregate is mixed with the binder. The alkali activator is mixed to the dry mixture of binders and fine aggregate. The mixing operation is continued for 4 min to get uniform colour and consistency. The fresh mix is shown in Fig. 1.

The fresh mix is placed in moulds and compacted. The specimens are demoulded after 12 h of adding activator to binder and cured in ambient condition. The room temperature was recorded and found to be between 25 and 34 °C. The relative humidity in the room was found to be between 65 and 80%. The specimen after curing was tested using 3000 kN digital compressive strength testing machine. The six standard mortar cubes of 50 cm^2 were prepared for all the 81 mixes given in Table 1. Out of six cubes, three were tested on 7 days and remaining three were tested on 28 days. The specimens were cured in ambient laboratory conditions and are given in Fig. 2.

Table 1 Details of geopolymer mortar mix and test results

Sl. No.	Designation of mortar mix	Weight of constituent materials in kg for cubic metre of mortar					Cube compressive strength in MPa at the age of	
		Fly ash	OPC	Sodium silicate solution	Sodium hydroxide solution	Fine aggregate	7 days ($f_{c\text{-}7d}$)	28 days ($f_{c\text{-}28d}$)
1	6/8/2.2/0.40	686	44	201	91	1178	34.9	62.3
2	6/8/2.2/0.45	686	44	226	103	1142	30.7	48.7
3	6/8/2.2/0.50	686	44	246	114	1105	35.6	67.0
4	6/8/2.5/0.40	686	44	209	83	1178	53.6	62.5
5	6/8/2.5/0.45	686	44	235	94	1142	43.5	76.1
6	6/8/2.5/0.50	686	44	261	104	1105	44.3	80.9
7	6/8/2.8/0.40	686	44	215	77	1178	54.6	88.1
8	6/8/2.8/0.45	686	44	242	86	1142	50.4	87.0
9	6/8/2.8/0.50	686	44	267	96	1105	37.5	65.9
10	6/12/2.2/0.40	686	44	201	91	1178	37.9	60.8
11	6/12/2.2/0.45	686	44	226	103	1142	44.8	67.0
12	6/12/2.2/0.50	686	44	246	114	1105	38.7	66.0
13	6/12/2.5/0.40	686	44	209	83	1178	60.4	86.5
14	6/12/2.5/0.45	686	44	235	94	1142	51.8	88.2
15	6/12/2.5/0.50	686	44	261	104	1105	52.2	82.8
16	6/12/2.8/0.40	686	44	215	77	1178	54.4	76.9
17	6/12/2.8/0.45	686	44	242	86	1142	54.0	105.5
18	6/12/2.8/0.50	686	44	267	96	1105	46.6	92.4
19	6/16/2.2/0.40	686	44	201	91	1178	45.8	79.9
20	6/16/2.2/0.45	686	44	226	103	1142	46.0	86.9
21	6/16/2.2/0.50	686	44	246	114	1105	40.1	74.2
22	6/16/2.5/0.40	686	44	209	83	1178	56.8	84.9
23	6/16/2.5/0.45	686	44	235	94	1142	47.5	93.5
24	6/16/2.5/0.50	686	44	261	104	1105	48.2	91.2
25	6/16/2.8/0.40	686	44	215	77	1178	54.0	95.8
26	6/16/2.8/0.45	686	44	242	86	1142	48.6	60.8
27	6/16/2.8/0.50	686	44	267	96	1105	45.2	48.0
28	9/8/2.2/0.40	664	66	201	91	1178	50.2	61.5
29	9/8/2.2/0.45	664	66	226	103	1142	53.4	69.3
30	9/8/2.2/0.50	664	66	246	114	1105	54.1	60.9
31	9/8/2.5/0.40	664	66	209	83	1178	55.1	76.5
32	9/8/2.5/0.45	664	66	235	94	1142	52.1	84.5
33	9/8/2.5/0.50	664	66	261	104	1105	48.0	90.1

(continued)

Table 1 (continued)

Sl. No.	Designation of mortar mix	Weight of constituent materials in kg for cubic metre of mortar					Cube compressive strength in MPa at the age of	
		Fly ash	OPC	Sodium silicate solution	Sodium hydroxide solution	Fine aggregate	7 days (f_{c-7d})	28 days (f_{c-28d})
34	9/8/2.8/0.40	664	66	215	77	1178	51.6	82.7
35	9/8/2.8/0.45	664	66	242	86	1142	58.4	74.5
36	9/8/2.8/0.50	664	66	267	96	1105	51.7	69.1
37	9/12/2.2/0.40	664	66	201	91	1178	47.7	60.7
38	9/12/2.2/0.45	664	66	226	103	1142	51.4	76.1
39	9/12/2.2/0.50	664	66	246	114	1105	44.6	71.9
40	9/12/2.5/0.40	664	66	209	83	1178	59.6	96.4
41	9/12/2.5/0.45	664	66	235	94	1142	42.8	86.4
42	9/12/2.5/0.50	664	66	261	104	1105	58.5	82.3
43	9/12/2.8/0.40	664	66	215	77	1178	56.4	84.5
44	9/12/2.8/0.45	664	66	242	86	1142	54.0	97.1
45	9/12/2.8/0.50	664	66	267	96	1105	49.8	82.8
46	9/16/2.2/0.40	664	66	201	91	1178	56.4	58.9
47	9/16/2.2/0.45	664	66	226	103	1142	55.1	81.1
48	9/16/2.2/0.50	664	66	246	114	1105	44.2	84.2
49	9/16/2.5/0.40	664	66	209	83	1178	56.4	78.9
50	9/16/2.5/0.45	664	66	235	94	1142	49.8	99.8
51	9/16/2.5/0.50	664	66	261	104	1105	45.2	89.2
52	9/16/2.8/0.40	664	66	215	77	1178	54.0	95.8
53	9/16/2.8/0.45	664	66	242	86	1142	48.6	60.8
54	9/16/2.8/0.50	664	66	267	96	1105	45.1	48.0
55	12/8/2.2/0.40	642	88	201	91	1178	48.4	62.4
56	12/8/2.2/0.45	642	88	226	103	1142	51.2	66.3
57	12/8/2.2/0.50	642	88	246	114	1105	45.5	73.34
58	12/8/2.5/0.40	642	88	209	83	1178	66.7	79.8
59	12/8/2.5/0.45	642	88	235	94	1142	48.3	50.2
60	12/8/2.5/0.50	642	88	261	104	1105	48.8	61.1
61	12/8/2.8/0.40	642	88	215	77	1178	64.4	75.8
62	12/8/2.8/0.45	642	88	242	86	1142	69.6	89.3
63	12/8/2.8/0.50	642	88	267	96	1105	66.8	85.1
64	12/12/2.2/0.40	642	88	201	91	1178	73.2	95.1
65	12/12/2.2/0.45	642	88	226	103	1142	52.9	86.8

(continued)

Table 1 (continued)

Sl. No.	Designation of mortar mix	Weight of constituent materials in kg for cubic metre of mortar					Cube compressive strength in MPa at the age of	
		Fly ash	OPC	Sodium silicate solution	Sodium hydroxide solution	Fine aggregate	7 days (f_{c-7d})	28 days (f_{c-28d})
66	12/12/2.2/0.50	642	88	246	114	1105	63.3	93.9
67	12/12/2.5/0.40	642	88	209	83	1178	38.0	39.1
68	12/12/2.5/0.45	642	88	235	94	1142	64.7	86.2
69	12/12/2.5/0.50	642	88	261	104	1105	54.7	64.4
70	12/12/2.8/0.40	642	88	215	77	1178	63.6	84.6
71	12/12/2.8/0.45	642	88	242	86	1142	66.9	67.8
72	12/12/2.8/0.50	642	88	267	96	1105	67.6	74.2
73	12/16/2.2/0.40	642	88	201	91	1178	74.9	92.9
74	12/16/2.2/0.45	642	88	226	103	1142	74.3	90.3
75	12/16/2.2/0.50	642	88	246	114	1105	71.1	73.4
76	12/16/2.5/0.40	642	88	209	83	1178	65.0	105.1
77	12/16/2.5/0.45	642	88	235	94	1142	65.6	104.2
78	12/16/2.5/0.50	642	88	261	104	1105	65.2	71.7
79	12/16/2.8/0.40	642	88	215	77	1178	55.1	69.7
80	12/16/2.8/0.45	642	88	242	86	1142	76.4	76.8
81	12/16/2.8/0.50	642	88	267	96	1105	58.7	79.1

Fig. 1 Appearance of fresh OPC included geopolymer mortar

3 Results and Discussions

The compressive strength of geopolymer mortar was determined on 7 and 28 days by laboratory test and is given in Table 1. The values given in Table 1 are the average of 3 cubes. The 7 days strength of geopolymer mortar was found to be between 30.7

Fig. 2 Ambient air curing of the specimens

and 76.4 MPa, and similarly, the 28 days strength was found to be between 39.1 and 105.5 MPa. This indicates that high-strength geopolymer mortar can be prepared with the addition of small quantity of OPC in conventional geopolymer mixture. The addition of calcium will significantly accelerate the setting and hardening of geopolymerisation slurries [8, 15]. It may be due to the precipitation of $Ca(OH)_2$ or calcium silicate hydrates, which triggers the geopolymer gel formation. Moreover, instead of heat curing which was used for conventional geopolymer concrete, ambient exposure curing is suggested for OPC included geopolymer concrete. Hence, it is expected that the OPC included geopolymer mortar is a viable composite material for in-situ construction. All specimens failed in shear, and a typical failure pattern is shown in Fig. 3.

The conical failure may be the interaction between frictional force mobilised at the specimen surface due to the platen of testing machine and lateral bulging force developed in the specimen. This is a typical failure pattern in ordinary concrete specimen, which indicates that the force transfer and redistribution in OPC included geopolymer mortar as similar to that of conventional concrete.

Fig. 3 Typical failure pattern in mortar cube specimen

4 Proposed Strength Prediction Models

Multiple regression analysis is used for developing a prediction model for the compressive strength of OPC included geopolymer mortar. The cement content (C), molar concentration (M), SS/SH ratio (S) and alkali–binder ratio (B) are considered as independent variable. The correlation coefficient is determined for C, M, S and B using the multiple regression analysis of the test data. The prediction models for 7 and 28 days compressive strength of mortar are given in Eqs. (1) and (2), respectively.

$$f_{c-7d} = 22.40 + 2.49C + 0.57M + 8.49S - 43.56B \quad (1)$$

$$f_{c-28d} = 53.87 + 0.12C + 1.04M + 9.07S - 27.71B \quad (2)$$

5 Comparison of Prediction with Experiment Data

The predicted strength of OPC included geopolymer mortar is compared with the corresponding experimental data given in Fig. 4. The deviation of predicted strength was found to be ±20% of experimental data of test results when compared to predicted results. A few of samples showed the some deviation from the above range. It may be due to the mixing process, compaction, etc., in the laboratory. The predicted strength is found to be comparable with the experimental data.

The test result data subjected to a comparison between predicted and experimental values is summarised in Table 2. It may be noted that the predicted strength is found to be in good agreement with the corresponding experimental data. The variation in the average predicted strength when compared to experimental data for the study by Nath and Sarkar [10] can be attributed to the variation in the chemical content of the constituents, ambient temperature, humidity conditions and machineries used for the preparation and testing speed. The proposed model predicted the test results of Saengsuree et al. [11] quite accurately.

The variation of the predicted compressive strength on 7 and 28 day is given in Figs. 5 and 6, respectively. The predicted 7d compressive strength was found to vary from 38 to 68 MPa. When the cement content is 6% and molarity of sodium hydroxide is 8, the predicted compressive strength was found to vary between 38 and 49 MPa. Similarly, when the cement content is 6% and molarity of sodium hydroxide is 8, the predicted compressive strength was found to vary between 58 and 68 MPa.

The predicted strength is found to be increasing with increase in the cement content (C) and molarity of sodium hydroxide (M). Similarly, the predicted strength increases with increase in the ratio of sodium silicate to sodium hydroxide (S). The predicted strength decreases with increase in the alkali–binder ratio (B). The 28-day strength predicted is higher than that of the 7-day strength.

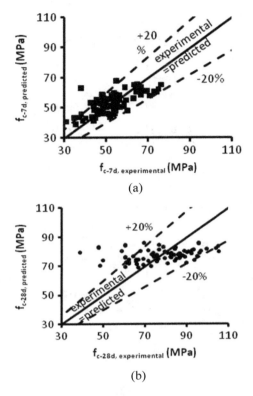

Fig. 4 Comparison of prediction with experimental data of the present study. **a** f_{c-7d}, **b** f_{c-28d}

6 Conclusion

In this investigation, an attempt was made to study the potential of using OPC included geopolymer mortar. The compressive strength at 7 and 28 days was determined after curing in ambient condition. The following conclusions are arrived at based on this study.

- By inclusion of a small percentage OPC to the conventional geopolymer mortar, 28-d compressive strength of 39–105 MPa was achieved.
- The heat curing can be completely avoided by inclusion of small quantity of OPC combined in geopolymer mix.
- The side face demoulding of specimens can be allowed at the age of 1 day for the case of OPC included geopolymer.

The mortar produced by low calcium fly ash and alkali activator with the inclusion of a small amount OPC can provide a feasible alternative to products produced by Portland cement. The prediction model of compressive strength of geopolymer mortar developed in this study can be used for ascertaining the strength capacity of

Table 2 Comparison of predicted strength with experimental data

Sl. No.	References	Range of variables				Property	Number of specimens[a]	Ratio of predicted to experimental	
		C	M	S	B			Average	SD[b]
1	Present study	6–12	8–16	2.2–2.8	0.40–0.50	f_{c-7d}	81 × 3 = 243	1.07	0.14
						f_{c-28d}	81 × 3 = 243	1.04	0.22
2	Nath and Sarkar [10]	5–12	14	1.5–2.5	0.35–0.45	f_{c-7d}	8 × 3 = 24	1.77	0.16
						f_{c-28d}	8 × 3 = 24	1.41	0.17
3	Saengsuree et al. [11]	5–15	10	0.67	0.40	f_{c-7d}	6 × 3 = 18	0.91	0.24
						f_{c-28d}	6 × 3 = 18	1.03	0.11

[a](Number of mixes) × (number of specimens in each mix)
[b]SD = standard deviation

Fig. 5 Variation of predicted 7-day compressive strength ($f_{c\text{-}7d}$)

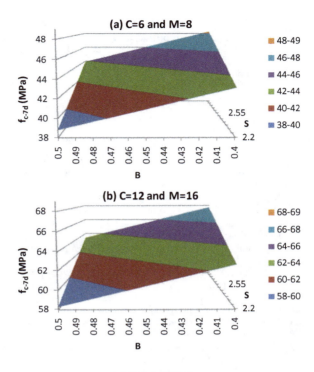

Fig. 6 Variation of predicted 28-day compressive strength ($f_{c\text{-}7d}$)

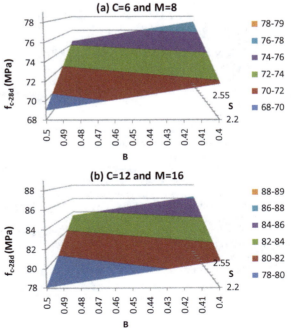

the geopolymer mortar. The advantage of the OPC blended geopolymer mortar is that no water curing is required and strength of the mixture is as high as 100 MPa. It is expected that the OPC included geopolymer mortar is a future material for repair works.

References

1. Davidovits, J. (1994). Properties of geopolymer cements. Paper presented at the First International Conference on Alkaline Cements and Concretes, Kiev State Technical University, Kiev, Ukraine.
2. Dombrowski, K., Buchwald, A., & Weil, M. (2007). The influence of calcium content on the structure and thermal performance of fly ash based geopolymers. *Journal of Materials Science, 42*(9), 3033–3043.
3. Granizo. M. L., Alonso, S., Blanco-Varela. M. T., & Palomo A. (2002). Alkaline activation of metakaolin effect of calcium hydroxide in the products of reaction. *Journal of the American Ceramic Society, 85,* 225–231.
4. IS 12269-1987. *Specification for 53 grade ordinary port land cement.* New Delhi: Bureau of Indian Standards.
5. IS 383-1970. *Specification for coarse and fine aggregates from natural sources for concrete.* sNew Delhi: Bureau of Indian Standards.
6. IS 3812-1981. *Specification for fly ash for use as pozzolana and admixture.* New Delhi: Bureau of Indian Standards.
7. Ken, P. W., Raml, M., & Ban, C. C. (2015). An overview on the influence of various factors on the properties of geopolymer concrete derived from industrial by-products. *Construction and Building Materials, 77,* 370–395.
8. Lee, W. K. W., & van Deventer, J. S. J. (2002). The effect of ionic contaminants on the early-age properties of alkali-activated fly ash-based cements. *Cement and Concrete Research, 32*(4), 577–584.
9. Nath, P., & Sarker, P. K. (2014). Effect of GGBFS on setting, workability and early strength properties of fly ash geopolymer concrete cured in ambient condition. *Construction and Building Materials, 66,* 163–171.
10. Nath, P., & Sarker, P. K. (2015). Use of OPC to improve setting and early strength properties of low calcium fly ash geopolymer concrete cured at room temperature. *Construction and Building Materials, 55,* 205–214.
11. Saengsuree, P., Tanakorn, P.-N., Sata, V., & Chindaprasirt, P. (2014). Influence of curing conditions on properties of high calcium fly ash geopolymer containing Portland cement as additive. *Materials and Design, 53,* 1906–1910.
12. Rangan, B. V. (2008). Fly ash-based geopolymer concrete. *Research report GC 4,* Perth, Australia: Curtin University of technology.
13. Rashad, A. (2014). A comprehensive overview about the influence of different admixtures and additives on the properties of alkali-activated fly ash. *Materials and Design, 53,* 1005–1025.
14. Somna, K., Jaturapitakkul, C., Kajitvichyanukul, P., & Chindaprasirt, P. (2011). NaOH-activated ground fly ash geopolymer cured at ambient temperature. *Fuel, 90*(6), 2118–2124.
15. Somna, K., & Bumrongjaroen. W. (2011). Effect of external and internal calcium in fly ash on geopolymer formation. In *Proceedings of 35th International Conference on Advanced Ceramics and Composites,* Daytona Beach, Florida, January.
16. Yip, C. K., Lukey, G. C., Provis, J. L., & van Deventer, J. S. J. (2008). Effect of calcium silicate sources on geopolymerization. *Cement and Concrete Research, 38,* 554–564.

Development of High Strength Lightweight Coconut Shell Aggregate Concrete

A. Sujatha and Deepa Balakrishnan

Abstract Coconut shell is an agricultural solid waste originating from the coconut industry. Coconut shell is used for many useful purposes but most of the coconut shell wastes are yet to be utilized commercially. Since coconut shells have the potential to be used as coarse aggregate in concrete, utilizing this waste in the construction industry not only reduces the solid waste management problems associated with it but also will be a valuable contribution to the industry as an eco-friendly construction material. In this investigation, coconut shell was used for the production of high strength lightweight concrete. The dry density and cube compressive strengths were measured. Coconut shell aggregate concrete of compressive strength 35.09 N/mm^2 and dry density 1913 kg/m^3 was produced. These are in the range of high strength lightweight aggregate concrete.

Keywords Lightweight aggregate · Coconut shell · High strength lightweight concrete

1 Introduction

Concrete made with natural aggregate has a low strength–weight ratio compared to steel. Hence, there is an economic disadvantage when we use this concrete for designing structural members for tall buildings, long span bridges and floating structures [1]. Main solution for this problem is the usage of high strength lightweight concrete. High strength low density concrete has strength levels in the range of 34–69 MPa, and its air dry density should not exceed 2000 kg/m^3 [2] Most popular method for the production of lightweight concrete(LWC) is to use lightweight aggregate [3]. In most of the cases, the lightweight concrete has been made using a

A. Sujatha (✉) · Deepa Balakrishnan
Cochin University of Science and Technology, Cochin, Kerala 682022, India
e-mail: a.sujatha89@gmail.com

Deepa Balakrishnan
e-mail: deepa_balu@cusat.ac.in

lightweight coarse aggregate and a normal weight fine aggregate [4]. Industrial wastes as well as other natural materials are used for the production of lightweight aggregate but their continuous extraction leads to depletion of their sources. These include aggregates prepared by expanding, pelletizing or sintering products such as blast furnace slag, clay, fly ash, shale, or slate and aggregates prepared by processing natural materials such as pumice, scoria or tuff [5]. Artificial aggregates such as expanded clay and shale are the most suitable lightweight aggregates for production of good-quality LWC [1] but these types of aggregates are produced through very high heat treatment process which results in high fuel costs [6]. Hence, it is necessary to find alternative sources for lightweight aggregate. In this context, agricultural wastes such as oil palm shell and coconut shell may be used for the production of lightweight aggregate concrete.

Coconut shell (CS) is a waste product from the agricultural industry and is available in plenty throughout the tropical regions of the world. India is the third-largest producer of coconut in the world (Fig. 1) [7]. Coconut shells are used for many purposes in different forms like powder, charcoal and activated carbon (Fig. 2.). (But most of the coconut shell wastes are yet to be utilized commercially. In many countries, the waste coconut shells are subjected to burning which liberates CO_2 into the atmosphere [8]. A solution to the challenges in coconut waste management may be achieved by using coconut shell as aggregate in concrete.

Previous studies on coconut shells have proven that CS possess superior qualities in terms of impact resistance, crushing strength and abrasion resistance compared to other conventional aggregates [9]. CS is different from other agricultural wastes since it contains high lignin and low cellulose. These properties results in high weather resistance and less moisture absorption than other agricultural wastes [10]. Sugar present in coconut shell is not in free form; hence, it could not affect the setting and

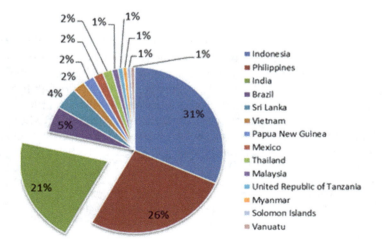

Fig. 1 World coconut production in 2016 [7]

Fig. 2 Coconut shell-based products [8]

strength attainment of concrete [10]. Hence, coconut shells can be used as coarse aggregate for production of structural lightweight concrete [11–14]. Water absorption of coconut shell aggregate concrete (CSAC) was more compared to other lightweight concrete but these property can be improved by proper curing and seasoning of coconut shells before using as aggregate [15]. Sorptivity of CSAC is comparable to other LWC [15, 16]. Other durability properties such as rapid chloride penetration and behaviour under elevated temperature were comparable to other LWC [15].

CSAC has been used in the production of non-structural elements such as non-pressure pipe, flooring tiles, paver bocks and manhole cover slabs of specified strength and durability. Although CS lightweight aggregate concrete has been successfully produced in the past, the compressive strength was generally in the range of 20–30 MPa. Not much has been reported on studies to develop high strength lightweight concrete by using this agricultural solid waste. Oil palm shells and coconut shells are similar in case of their properties such as bulk density and specific gravity [11]. Oil palm shells (OPS) are used for production of high strength lightweight concrete (HSLWC) so that there is a scope to develop high strength lightweight concrete using coconut shell aggregates [17]. So the main purpose of this study is to develop high strength lightweight concrete using coconut shell as coarse aggregate.

2 Experimental Programme

2.1 *Materials Used*

2.1.1 Cement

Cement is the binding material, which binds the individual units of aggregates into solid mass by virtue of its inherent properties of setting and hardening in combination with water. OPC 53 grade cement was used for experimental study.

2.1.2 Coconut Shell Aggregates

Coconut shells were collected from nearby oil mills. The shells were then soaked in water for one week for seasoning as mentioned by Gunasekaran et al. [13]. Then the hardened surface of coconut shell was cleaned to remove fibres, mud, etc. The coconut shells were first crushed manually using hammer, and then they were crushed to required size in coconut shell crushing machine. The coconut shell crushing machine was specially fabricated for this purpose only (Fig. 3a). Coconut shells and crushed coconut shells are shown in Fig. 3b, c. Properties of CS are given in Table 1.

Fig. 3 **a** Coconut shell crusher. **b** Coconut shells. **c** Crushed coconut shells

Table 1 Properties of aggregates

S. No.	Test conducted	Fine aggregate	CS aggregate
1	Specific gravity	2.40	1.15
2	Bulk density (kg/m^3)	1650	665
3	Fineness modulus	2.94	6.78
4	Maximum size (mm)	4.76	10

2.1.3 M Sand

M sand was used as fine aggregate. Tests on fine aggregate were carried out as per IS 2386-1963(Part III) to find the physical properties. Sieve analysis was conducted to determine the fineness modulus and particle size distribution as given in Table 1.

2.1.4 Super Plasticizer

Polycarboxylate ether-based water reducing admixture cera hyperplastic XR-W40 was used as super plasticizer for the production of coconut shell aggregate concrete. Super plasticizer used for the study was 1% of cement weight for all the mixes.

2.2 Mix Proportioning

No standard methods are available for mix design of CSAC since the coconut shells are wood-like materials and different from natural aggregates. High strength lightweight concrete with oil palm shells as coarse aggregate was developed by Shafigh et al. [18]. The properties of oil palm shell and coconut shell are comparable [13]. Hence, it is also possible to develop CS HSLWC. As a trial, the mix proportions developed by Shafigh et al. [18] for HSLWC oil palm shell concrete were adopted for the study. Different cement contents of 480, 520 and 550 (kg/m^3) were considered. Maximum size of CS aggregate used was 10 mm. Total 27 cubes, nine for each mix were cast and tested (Fig. 4). Compressive strength as well as dry density at 3, 7 and 28 days were determined. The mix proportions used are given in Table 2.

Fig. 4 Specimens casting and testing

Table 2 Mix proportions

Mix	Cement (C) (kg/m^3)	Fine aggregate (FA) (kg/m^3)	Coconut shell (CS) (kg/m^3)	Water(W) (kg/m^3)	Mix ratio (C:FA:CS:W)	Average 28 day dry density (kg/m^3)
M1	480	1050	295	182	1:2.18:0.62:0.38	1928
M2	520	858	364	197.6	1:1.65:0.7:0.38	1958
M3	550	950	273	234	1:1.72:0.496:0.425	1913

3 Results and Discussions

All the three mixes exhibited a good high workability. No segregation and bleeding were occurred in any of the mixes. As per Shafigh et al. [18] by improving the workability and compatibility of the mixture, the strength of the interfacial transition zone tends to improve and thus improves the strength of the concrete.

The 28-day compressive strength of the CS aggregate concretes varies between around 33 and 35 MPa. The dry densities of all three mixes were below 2000 kg/m^3. Maximum compressive strength of 35.09 MPa and average dry density of 1913 kg/m^3 were obtained for M3 mix. High strength low density concrete has strength levels in the range of 34–69 MPa, and its air dry density should not exceed 2000 kg/m^3[2]. Hence, M3 mix can be designated as high strength lightweight concrete.

CS–cement ratios of mixes M1, M2 and M3 are 0.62, 0.70 and 0.496, respectively. As per Geiver et al.[17], wood–cement ratio below 0.5 had an adverse effect on strength of cement concrete composites, and this may be the reason that even with increased cement content there was not much increase in strength of M3 as compared to mixes M1 and M2. As mentioned by Gunasekaran et al. [13], wood–cement ratio has strong influence on strength attainment of concrete [13]. Since coconut shells are wood-like material, we should consider CS–cement ratio along with water–cement ratio for further trial mix design of CS HSLWC.

Comparison of compressive strength variation with respect to age of CSAC mixes and oil palm shell concrete (OPSC) mixes is shown in Figs. 5, 6 and 7. The strengths of CSAC mixes were lower than that of OPSC mixes. This variation may be due to many factors such as quality of materials used, environmental conditions and biological properties of CS and OPS aggregates.

4 Conclusion

An attempt was done to develop high strength lightweight aggregate concrete by using crushed coconut shells as coarse aggregate. From the study, following conclusions were drawn.

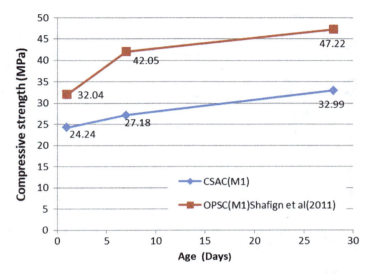

Fig. 5 Comparison of strength variation with respect to age of CSAC (M1) and OPSC of same mix ratio

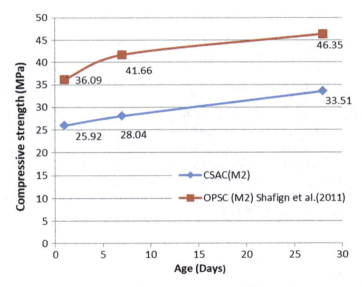

Fig. 6 Comparison of strength variation with respect to age of CSAC (M2) and OPSC of same mix ratio

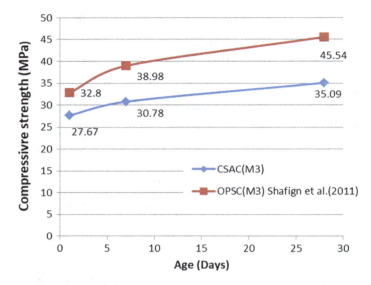

Fig. 7 Comparison of strength variation with respect to age of CSAC (M3) and OPSC of same mix ratio

- Maximum 28-day compressive strength of 35.09 MPa and dry density of 1913 kg/m^3 were obtained for M3 mix. As per Hoff, G. C. (2002) high strength low density concrete has strength levels in the range of 34 to 69 MPa, and its air dry density should not exceed 2000 kg/m^3. Hence, M3 mix is designated as high strength lightweight concrete.
- From the present study, it can be concluded that CS has the potential to be used as lightweight aggregate for developing high strength lightweight concrete.
- Further trial mixing has to be done to improve the strength of coconut shell aggregate concrete by considering CS–cement ratio along with other factors.
- High amount of cement content is essential for production of high strength lightweight concrete. Hence, further studies considering mineral admixtures as replacement for cement are recommended.

References

1. Mehta, P. K., & Monteiro, P. J. M. (2006). *Concrete microstructure, properties, and materials* (3rd ed.). New York: McGraw-Hill.
2. Hoff, G. C. (2002). *Guide for the use of low-density concrete in civil works projects*. Tech. Rep. ERDC/GSL TR-02-13 (TR INP-02-7), US Army Corps of Engineers, Engineer Research and Development Center.
3. Polat, R., Demirboğa, R., Karakoç, M. B., & Türkmen, İ. (2010). The influence of lightweight aggregate on the physico-mechanical properties of concrete exposed to freeze–thaw cycles. *Cold Regions Science and Technology, 60*(1), 51–56.

4. Boyd, S. R., Holm, T. A., & Bremner, T. W. (2006). Performance of structural lightweight concrete made with a potentially reactive natural sand. *Special Publication, 234,* 737–746.
5. ASTM, C. (2006). 330-05. *Standard specification for lightweight aggregates for structural concrete.* West Conshohocken, PA: ASTM International.
6. Zhang, M. H., & Gjorv, O. E. (1991). Characteristics of lightweight aggregates for high-strength concrete. *Materials Journal, 88*(2), 150–158.
7. https://www.worldatlas.com/articles/the-world-leaders-in-coconut-production.html.
8. https://coconutboard.nic.in/CoconutProducts.aspx#CoconutShellBasedProducts.
9. Olanipekun, E. A., Olusola, K. O., & Ata, O. (2005). A comparative study of concrete properties using coconut shell and palm kernel shell as coarse aggregates. *Building and Environment, 4,* 297–301.
10. Gunasekaran, K., Kumar, P. S., & Lakshmipathy, M. (2011a). Study on properties of coconut shell as an aggregate for concrete. *Indian Concrete Institute Journal, 12*(2), 27–33.
11. Jaya Prithika, A., & Sekar, S. K. (2016a). Mechanical and fracture characteristics of Eco-friendly concrete produced using coconut shell, ground granulated blast furnace slag and manufactured sand. *Construction and Building Materials, 103,* 1–7.
12. Kanojia, A., & Jain, S. K. (2017). Performance of coconut shell as coarse aggregate in concrete. *Construction and Building Materials, 140,* 150–156.
13. Gunasekaran, K., Kumar, P. S., & Lakshmipathy, M. (2011b). Mechanical and bond properties of coconut shell concrete. *Construction and Building Materials., 25,* 92–97.
14. Nadir, Y., & Sujatha, A. (2018a). Bond strength determination between coconut shell aggregate concrete and steel reinforcement by pull-out test. *Asian Journal of Civil Engineering, 19*(6), 713–723.
15. Gunasekaran, K., Annadurai, R., & Kumar, P. S. (2015). A study on some durability properties of coconut shell aggregate concrete. *Materials and Structures, 48,* 1253–1264.
16. Nadir, Y., & Sujatha, A. (2018b). Durability properties of coconut shell aggregate concrete. *KSCE Journal of Civil Engineering, 22*(5), 1920–1926.
17. Geiver, R. L., Souza, M. R., Moslerni, A. A., & Sirnatupang, M. H. (1992). Carbon dioxide application for rapid production of cement particleboard. In *Proceedings, Inorganic-Bonded Wood and Fiber Composite Materials* (Vol. 3, pp. 31–41). Madison, Wis: Forest Prod Soc.
18. Shafigh, P., Jumaat, Z. M., & Mahmud, H. (2011). Oil palm shell as a lightweight aggregate for production high strength lightweight concrete. *Construction and Building Materials, 25*(4), 1848–1853.

Comparison of the Performance Between Concrete Filled and Stiffened LDSS Column

Divya Roy and Milu Mary Jacob

Abstract Stainless steel is employed in a wide range of structural applications such as in bridges, storage tanks and reinforcing bars for concrete structures. Among the various grades of stainless steel, austenitic grades are most popular in the construction industry which has nickel content 8–11%. Recently, a new form of a duplex stainless steel is developed which is lean duplex stainless steel (LDSS), which has a lower nickel content of about 1.5%. This grade has lower-cost, improved corrosion resistance and strength, enabling a reduction in section sizes leading to higher strength to weight ratios. The particular grade used in this study is EN 1.4162, which is generally less expensive than the austenitic counterpart but offers higher strength and a reasonable corrosion resistance. In this study, the buckling performance of different shaped (square, L, T, cross (+)) lean duplex stainless steel hollow stub columns with stiffeners are investigated. Also performance of LDSS stub column is compared with concrete filled stub column. A nonlinear static analysis of LDSS stub column is studied using ANSYS workbench. In this project, it is found out that buckling capacity increases by changing the section from square-T-L–cross shape. It is also inferred that sections with stiffeners at the corners show better performance than straight stiffeners.

Keywords Hollow stub column · Buckling load · LDSS · Finite element analysis · Stiffeners · Nonlinear static analysis

D. Roy (✉) · M. M. Jacob
Department of Civil Engineering, Saintgits College of Engineering, Kottayam, Kerala, India
e-mail: divyaroseroy@gmail.com

M. M. Jacob
e-mail: milu.mary@saintgits.org

1 Introduction

1.1 General

The developments occurred in material processing, providing a range of stainless steel which can overcome all the drawbacks of carbon steel such as low corrosion resistance and higher material cost. The main advantages of stainless steel are higher corrosion resistance, high ductility, high strength, impact resistance, smooth and uniform surface, aesthetic appearances and ease of maintenance and construction [1]. Depending on the corrosion resistance and alloy contents, the duplex stainless steel grades can be classified into different type, lean duplex grade being one of them [2]. This has led to the widespread use of LDSS in construction industry. In this study, the performance of different shaped concrete filled lean duplex stainless steel hollow short column and hollow short column with stiffeners is analysed. Studies show that concrete confinement increases the compressive strength of the structure [3]. The results for concrete filled LDSS column and hollow column with stiffeners are compared. The selected shapes for analysis are square, L, T and + cross sections. Columns with these kinds of cross sections can be used as corner columns in framed structures. They also have the advantage in providing a flushed wall face, resulting in an enlarged usable indoor floor space area and also in making the interior space more regular. Modelling and analysis are done using ANSYS workbench software. Limited studies have been made on stub column with non-rectangular cross sections, as presented in the current research. The tubular structures are more prone to buckling and deformation due to its light and thin sections. An efficient and cost-effective way to improve the structure over these problems is to add stiffeners, which can substantially increase the buckling load capacity [4]. The FE parametric results of LDSS hollow stub columns, concrete filled and stiffened column are compared with concrete stub column.

2 Numerical Investigation

A total of 16 models are analysed in ANSYS software. The cross sections used are square, L, T and cross (+) sections. Under each section, concrete filled, hollow and stiffened column sections are analysed in order to compare the results. In the case of stiffened sections, two types of stiffening arrangements are considered, i.e. at the corners and at the straight sides of the sections and both are analysed. Incremental loading is provided at the top end of each section. Boundary condition adopted is fixed bottom end in all column sections. All the specimens are of depth (D) of 600 mm, breadth (B) of 600 mm, thickness (t) of 10 mm, length (L) of 1800 mm and width of stiffener is 80 mm. The stiffeners are provided throughout the length of the column.

Table 1 Material properties of test specimen

Section	E_0 (MPa)	σ_u (MPa)	ε_f (%)
S1	209,800	839	44
S2	199,900	773	42
S3	198,800	761	47

2.1 LDSS Material Properties

Properties of test specimen are shown in Table 1 [5].

2.2 Buckling Analysis

Structural members subjected to high compressive stress result in sideways sudden failures which are characterized as buckling. The ultimate compressive stress of the material is more than the compressive stress at the point of failure. The member becomes unstable when the applied load is increased which results in buckling. With further increase in load, unpredictable deformation occurs and leads to complete loss of load carrying capacity of member.

2.3 Modelling

The modelling of stub column is carried out in ANSYS. Numerical analysis is done to study the critical buckling load of each section using ANSYS software. The cross section was modelled using key points, lines and arcs. Further, cross-sectional shape was dragged along a line to generate the section. The maximum load applied is 15000 kN and was applied by incremental method. The section was developed using ANSYS element type SHELL181. The mapped meshing technique was used to generate a variable density mesh. Fine mesh was employed for the sections. The bottom end was restricted for six degree of freedoms to represent the fixed end boundary condition. Load was applied as nodal displacements at the upper end. The loading diagram for each column sections obtained from software is shown in Figs. 1, 2, 3 and 4.

3 Results and Discussions

The buckling shape and buckling load are shown in Figs. 5, 6, 7 and 8.

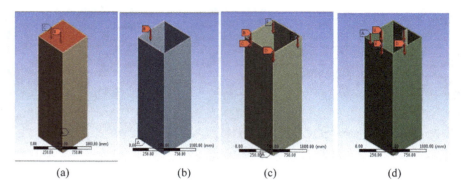

Fig. 1 Model—square section **a** Concrete filled **b** Hollow **c** Stiffeners at corners **d** Stiffeners at straight edges

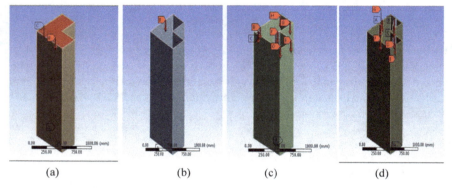

Fig. 2 Model—L section **a** Concrete filled **b** Hollow **c** Stiffeners at corners **d** Stiffeners at straight edges

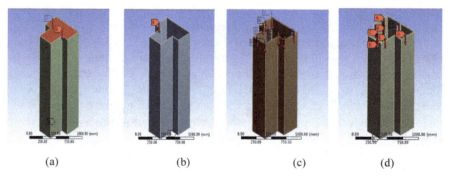

Fig. 3 Model—T section **a** Concrete filled **b** Hollow **c** Stiffeners at corners **d** Stiffeners at straight edges

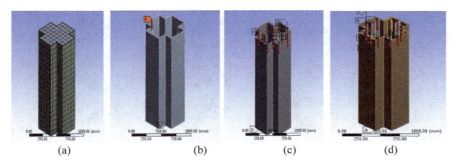

Fig. 4 Model—Cross section **a** Concrete filled **b** Hollow **c** Stiffeners at corners **d** Stiffeners at straight edges

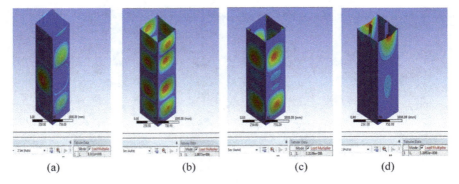

Fig. 5 Buckling shape—Square sections **a** Concrete filled **b** Hollow **c** Stiffeners at corners **d** Stiffeners at straight edges

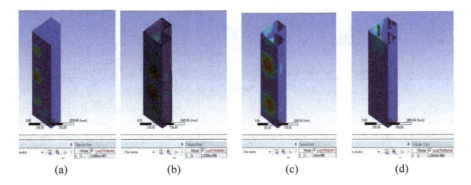

Fig. 6 Buckling shape—L sections **a** Concrete filled **b** Hollow **c** Stiffeners at corners **d** Stiffeners at straight edges

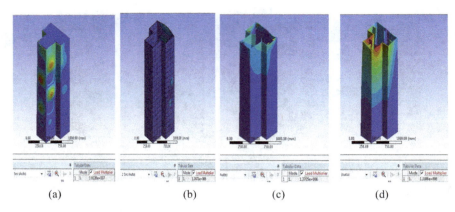

Fig. 7 Buckling shape—T sections **a** Concrete filled **b** Hollow **c** Stiffeners at corners **d** Stiffeners at straight edges

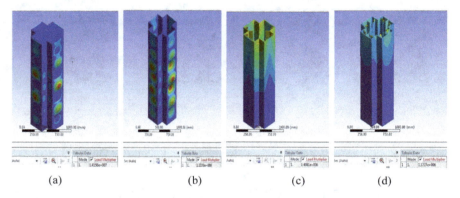

Fig. 8 Buckling shape—Cross sections **a** Concrete filled **b** Hollow **c** Stiffeners at corners **d** Stiffeners at straight edges

Table 2 Buckling capacities of different sections obtained after analysis

S. No.	Model	Concrete filled (kN)	Hollow (kN)	Stiffener—straight (kN)	Stiffener—corners (kN)
1	Square	8012	1087.7	1195.3	1310.9
2	T shape	10,226	1247.1	1330.6	1370.5
3	L shape	10,381	1270.5	1341.5	1410
4	Cross	14,156	1137.9	1199.7	1408.1

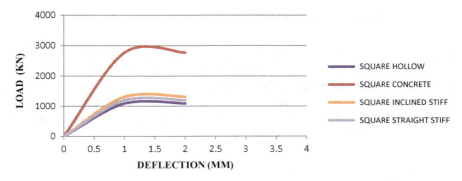

Fig. 9 Load deflection graph—Square section

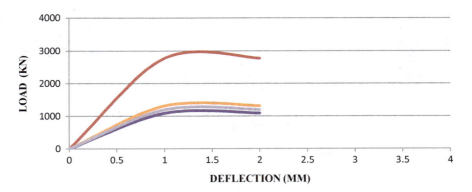

Fig. 10 Load deflection graph—T section

3.1 Buckling Capacity (kN)

The buckling capacities of different sections are summarized in Table 2. It can be concluded that the non-rectangular sections cross, L, T higher buckling load compared to the rectangular section. Buckling capacity increases as the sections change from square-T-L-cross shape. The highest buckling load capacity is shown by cross concrete filled section and the least by square concrete filled section. It can also be inferred that increasing the number of sides of the section increases the buckling load carrying capacity.

3.2 Load Deflection Graphs

The load deflection diagram of square, T, L, cross sections is plotted in Figs. 9, 10, 11 and 12. It can be inferred from the graphs that buckling capacity increases as the sections change from square-T-L-cross shape.

A comparison is made between the buckling capacities of different sections with respective concrete filled, hollow, hollow stiffened (inclined) and hollow stiffened (straight) sections. This is given in Table 3.

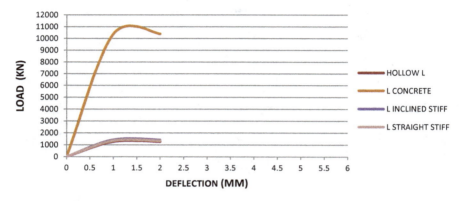

Fig. 11 Load deflection graph—L section

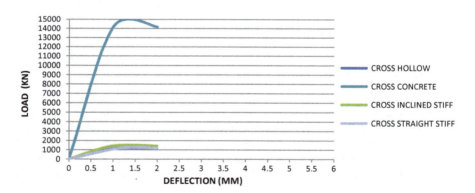

Fig. 12 Load deflection graph—Cross section

Table 3 Comparison with respective square sections—concrete filled, hollow, hollow stiffened (corners) and hollow stiffened (straight)

S. No.	Model	Increase in buckling load
1	Cross concrete filled	1.8
2	Cross hollow	1.05
3	Cross stiffened—corners	1.7
4	Cross stiffened—straight	1.003
5	L concrete filled	1.4
6	L hollow	1.16
7	L stiffened—corners	1.075
8	L stiffened—straight	1.122
9	T concrete filled	1.2
10	T hollow	1.14
11	T stiffened—corners	1.04
12	T stiffened—straight	1.13

4 Conclusion

Finite element models were developed to simulate buckling failure of LDSS column sections under compression. ANSYS finite element package was used for the analysis, and nonlinear material behaviour was studied in order to represent the plastic failure in the post-yield region. Results showed that the displacement of the structure under compression was reduced with the incorporation of stiffeners. FE analysis of non-rectangular sections and representative square sections with concrete filled, hollow and sections with stiffeners are compared. The non-rectangular sections cross, L, T higher buckling load are compared to square section. The percentage increases in buckling capacities are 27.6, 29.4, 76.48% for T, L and cross sections, respectively. Buckling capacity increases as the sections change from square-T-L-cross shape. Column with stiffeners given at the corners (inclined) showed better buckling. The buckling capacity increases were 9.6, 2.9, 5.1, 20% for square, T section, L section, cross sections, respectively.

Acknowledgements We acknowledge the support rendered by everyone.

References

1. Khate, K., Patton, M. L., & Marthong, C. (2018). Structural behaviour of stainless steel stub column under axial compression: a FE study, *International Journal of steel structures, 18*, 1723–1740. https://doi.org/10.1007/s13296-018-0083-1.
2. Huang, Y., & Young, B. (2012). Material properties of cold formed lean duplex stainless steel sections. *Thin Walled Structures, 54*, 72–81. https://doi.org/10.1016/j.tws.2012.02.003

3. Hassanein, M., Kharoob, O., & Liang, Q. (2015). Behaviour of circular concrete filled LDSS tubular short columns. *Thin Walled Structures, 68,* 113–123. https://doi.org/10.1016/j.tws.2013.03.013
4. Vetsa, G. S., & Singh, K. D (2015). Effect of stiffeners in lean duplex stainless steel (LDSS) hollow circular stub columns under pure axial compression. (*IJIRSET*), *3*(4)
5. Lam, D., Yang, J., & Dai, X. (2019). Finite element analysis of concrete filled lean duplex stainless steel column. *Structures, 21,* 150–155. https://doi.org/10.1016/j.istruc.2019.01.024

Aspect Ratio Factor for Strength Correction of Pressed Earth Brick Prisms

Nassif Nazeer Thaickavil and Job Thomas

Abstract This paper presents an experimental study considering the effect of aspect ratio of masonry prism specimens on the compressive strength of pressed earth brick masonry. The variable considered in the experimental study is the aspect ratio of prism specimens. A total of twelve specimens were prepared and tested under four different geometric configurations. Strength correction factors to account for the effects of aspect ratio of prism have been proposed by conducting a regression analysis.

Keywords Pressed earth brick masonry · Aspect ratio · Prism specimens · Strength correction factor

1 Introduction

Masonry construction is one of the oldest construction techniques known to mankind. Masonry typically consists of masonry units held together by a mortar. The masonry units and the mortar may be made of a wide variety of materials. The earliest known bricks were made of sun-burnt mud [1], and the first fired bricks discovered in Mesopotamia are at least 5500 years old [2]. The popular masonry units made of earth include adobe bricks, cob bricks, burnt clay bricks and pressed earth bricks [3–5].

Pressed earth bricks are a sustainable alternative to burnt clay bricks [6, 7] and are manufactured by compressing a mixture of soil, water, additives and stabilizers either manually or by mechanized means. These are also called as stabilized mud blocks or soil cement blocks. Stabilizers such as cement or lime are used during the manufacture of these bricks to improve the compressive strength and durability of these bricks. These bricks have the advantages of better cost effectiveness, fire resistance, low thermal conductivity, low porosity and better energy efficiency when compared to burnt clay bricks [6, 7]. The compressive strength of cement stabilized pressed earth

---

N. N. Thaickavil (✉) · J. Thomas
Department of Civil Engineering, Cochin University of Science and Technology, Kochi, Kerala 682022, India
e-mail: nassifnazeer@gmail.com

bricks depends on the compacting effort (density), cement content, moisture content and clay content [8]. The manufacture of pressed earth bricks being sustainable in nature offers a viable alternative for the construction of low-cost buildings in the country. IS: 1725 [9] is the governing standard for use of soil-based blocks in India.

Although masonry has been known to man for a long time, studies to understand its actual behavior have been conducted only in the last century or so. The compressive strength of masonry is the most important parameter that has to be known before designing masonry structures. Some codes and design standards suggest empirical formula to determine the masonry strength. Designers make use of such formula to design masonry structures. IS: 1905 [10] prescribes the basic compressive stress of masonry for only a limited set of brick and mortar combinations. For a more accurate determination of the compressive strength, experimental studies have to be conducted on specimens of masonry.

The compressive strength of masonry can be studied by testing specimens of masonry prisms in the laboratory. Prism specimens are small assemblages of masonry units and mortar usually one unit thick. Prisms are believed to be a better representation of the actual masonry construction when compared to testing the individual masonry units as the effects of the properties of the constituents and quality of workmanship are taken into account [11].

Stack-bonded configuration is the most commonly adopted method for experimental investigations [12]. IS: 1905 [10] prescribes that masonry prisms should have a minimum height of 40 cm and that the aspect ratio should be between 2 and 5. ASTM E447 [13] suggests that the minimum height of the prism specimen should be 38.1 cm. The compressive strength of cement stabilized pressed earth block prisms is found to decrease with the increase in aspect ratio as reported by Walker [14]. Bei and Papaiyanni [15] concluded that doublet and triplet specimens better represented the wall behavior.

The slenderness of prism specimens affects the values of compressive strength, and hence, strength correction factors have to be introduced to exclude the slenderness effects [8]. Various international codes propose discrete values for correction factors to account for the effects of differences in aspect ratio on the compressive strength of prisms [10, 16, 17]. Experimental investigations have to be carried out for understanding the exact behavior of pressed earth brick masonry as the research data available is scarce. Strength correction factors are proposed in this paper to account for the differences in the strength of prism specimens.

2 Experimental Program

Single wythe brickwork prism specimens, two to five units high, were made in stack bond arrangement for four different configurations as shown in Fig. 1. A total of 12 brick masonry prisms were prepared with three specimens for each of the four configurations.

Fig. 1 Configuration of prism specimens

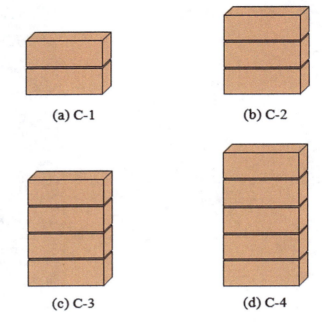

2.1 Materials

The constituent materials used to prepare masonry prisms were tested prior to the experimental program. Cement stabilized pressed earth bricks and cement–sand mortar of proportion 1:6 were the primary materials used in this study. The mean dimensions of the brick were found to be 190 mm × 113 mm × 100 mm, and its compressive strength (f_b) was found out to be 4.56 MPa. The mortar was prepared using 43-grade OPC cement and manufactured sand conforming to Zone II of IS: 383 [18]. The compressive strength test of mortar was determined by testing 70.7 mm cubes as per IS: 2250 [19]. The mean compressive strength of mortar (f_m) was obtained as 13.6 MPa.

2.2 Details of Specimens

To determine the compressive strength of masonry using pressed earth blocks and cement–sand mortar, prism specimens were prepared. A skilled mason was employed for the making of specimens, and the thickness of the bed joint mortar was maintained approximately as 10 mm for all joints using a template. All the specimens were cured for 28 days by spraying water at regular intervals.

Fig. 2 Testing of specimen C-2

2.3 Testing

After curing the prisms for a period of 28 days, all specimens were capped with a thin layer of dental plaster of thickness 1–2 before testing. This ensured a level contact surface between the platens of the testing machine. The load was applied in stages. One half of the expected load was applied initially at a convenient rate, and the remaining load was applied at a uniform rate extending two hours. Figure 2 shows the testing of prism specimen C-2 in the laboratory.

The ultimate load ($P_{u\text{-}e}$) was recorded, and the strength of the prism ($f_{p\text{-}e}$) is calculated from Eq. 1.

$$f_{p-e} = P_{u-e}/\text{Average cross-sectional area of specimen} \qquad (1)$$

2.4 Results

The results of the experiment have been tabulated in Table 1 showing specimen designation, height of prism specimen (h) and mean prism strength obtained from the experiment (f_{pe}) where h is in mm and f_{pe} is in MPa. Tall specimens failed in lateral tension with the formation of outward bursting cracks in the middle of the specimen, while shorter specimens failed in shear.

Table 1 Results of the experimental study

Specimen designation	H (mm)	$f_{p\text{-}e}$ (MPa)
C-1	210	1.27
C-2	320	1.00
C-3	430	0.88
C-4	540	0.76

2.5 Effect of Aspect Ratio on Masonry Strength

The prism strength decreases with the increase in the aspect ratio. This is in accordance with the observations of Hamid et al. [20]. The prism strength is affected by the effects of end platens due to the restraining effect caused by the mobilization of frictional forces between the top of the specimen and the end platens. The end zones of the prism are in compression while the mid zone is subjected to lateral tension. The area of tension zone is comparatively greater for specimens with greater height to thickness ratio, and hence, failure occurs by formation of vertical splitting cracks along the mid height. The strengths of prisms are thus affected by the slenderness of the specimens, and hence, correction factors have been proposed in this paper to account for this difference.

3 Strength Correction Factors

The strength of the prism can be estimated empirically by making use of the formulae available in various codes and design standards. The formula proposed by MSJC [21] to determine the compressive strength of solid clay brick masonry is represented by Eq. 2.

$$f'_m = A \times (400 + 145 \times B \times f_u)/145 \qquad (2)$$

where f'_m is the specified compressive strength of masonry in MPa, and f_u is the average compressive strength of brick in MPa. A is the quality assurance factor, and B is the mortar grade factor. The model proposed by Eq. 2 is used for predicting the strength of five course masonry prisms. The equation for predicting prism compressive strength can be modified by applying the slenderness correction factor and is given by Eq. 3.

$$f'_{m-p} = C \times f'_m \qquad (3)$$

where f'_{m-p} is the compressive strength corrected for slenderness in MPa, and C is the correction factor to incorporate the effects of h/t ratio.

Fig. 3 Influence of h/t ratio on C

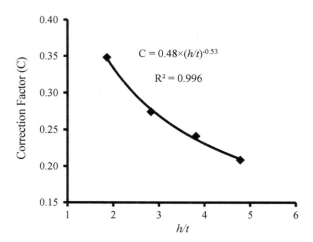

Correction factor as given in ASTM C 1314 [16] is considered to account for the effects of slenderness of the pressed earth brick masonry prism specimens. The variation of C with respect to the aspect ratio (h/t) is plotted in Fig. 3.

Based on the experimental data, a model for prediction of C is proposed by regression analysis and is given in Eq. 4.

$$C = 0.48 \times (h/t)^{-0.53} \qquad (4)$$

The proposed correction factors for the compressive strength of prisms based on Eq. 4 are given in Table 2. The correction factors so obtained were used to determine the predicted prism compressive strength ($f_{p\text{-}p}$), and the values are tabulated.

Figure 4 shows the comparison of the experimentally obtained prism compressive strength ($f_{p\text{-}e}$) and the predicted prism compressive strength ($f_{p\text{-}p}$). The proposed model predicted the strength of prisms accurately.

Table 2 Prism strengths based on proposed correction factors

Specimen designation	h/t	Correction factor	$f_{p\text{-}p}$ (MPa)
C-1	1.86	0.35	1.26
C-2	2.83	0.28	1.01
C-3	3.81	0.24	0.86
C-4	4.78	0.21	0.76

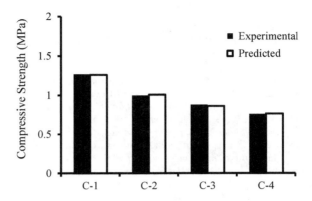

Fig. 4 Comparison of predicted and experimental strength

4 Conclusion

The results of the experimental study on pressed earth brick masonry show that the compressive strength of masonry prisms decreases with the increase in aspect ratio. A model has been proposed to account for the effects of slenderness on the compressive strength of prism specimens. The compressive strength values obtained from the experiment were compared with the predicted compressive strength of prisms using the MSJC model and the proposed correction factors. The proposed model is found to perform satisfactorily.

References

1. Fiala, J., Mikolas, M., & Krejsova, K. (2019). Full brick, history and future. In *IOP Conference Series: Earth and Environmental Science*, Vol. 221, Prague.
2. Hnaihen, K. H. (2020). The appearance of bricks in ancient mesopotamia. *Athens Journal of History, 6*(1), 73–96.
3. Fabbri, A., Morel, J. C., & Gallipoli, D. (2018). Assessing the performance of earth building materials: a review of recent developments. *RILEM Technical Letters, 3*, 46–58.
4. Quagliarini, E., Stazi, A., Pasqualini, E., & Fratalocchi, E. (2010). Cob construction in Italy: Some lessons from the past. *Sustainability, 2*, 3291–3308.
5. Maniatidis, V., & Walker, P. (2003). *A review of rammed earth construction*. Bath: University of Bath.
6. Malkanthi, S. N., Balthazaar, N., & Perera, A. A. D. A. J. (2020). Lime stabilization for compressed stabilized earth blocks with reduced clay and silt. *Case Studies in Construction Materials, 12*, 1–8.
7. Riza, F. V., Abdul Rahman, I., & Zaidi, A. M. A. (2010). A brief review of compressed stabilized earth brick (CSEB). In *2010 International Conference on Science and Social Research (CSSR 2010)*, December 5–7, 2010, Kuala Lumpur, Malaysia.
8. Morel, J., Pkla, A., & Walker, P. (2007). Compressive strength testing of compressed earth blocks. *Construction and Building Materials, 21*, 303–309.
9. IS: 1725. (2013). *Indian standard specification for soil based blocks used in general building construction*. Bureau of Indian Standards.

10. IS: 1905. (1987). *Indian standard code of practice for structural use of unreinforced masonry*. Bureau of Indian Standards.
11. Thaickavil, N. N., & Thomas, J. (2018). Behaviour and strength assessment of masonry prisms. *Case Studies in Construction Materials, 8*, 23–38.
12. Ganesan, T. P., & Ramamurthy, K. (1992). Behavior of concrete hollow-block masonry prisms under axial compression. *Journal of Structural Engineering, 118*, 1751–1769.
13. ASTM. E447. (1997). Test methods for compressive strength of laboratory constructed masonry prisms. *American Society for Testing and Materials*.
14. Walker, P. (1997). Characteristics of pressed earth blocks in compression. In *Proceedings of the 11th International Brick and Block Masonry Conference*, Shanghai, p. 1.
15. Bei, G., & Papayianni, I. (2003). Strength of compressed earth block masonry. *WIT Transactions on the Built Environment, 66*, 367–375.
16. ASTM C 1314. (2003). Standard test method for compressive strength of masonry prisms. *American Society for Testing and Materials*.
17. Aubert, J. E., Maillard, P., Morel, J., & Al Rafii, M. (2016). Towards a simple compressive strength test for earth bricks? *Materials and Structures, 49*(5), 1641–1654 (Springer Verlag).
18. IS: 383. (1970). *Indian standard specification for coarse and fine aggregates from natural sources for concrete*. Bureau of Indian Standards.
19. IS: 2250. (1981). *Indian standard code of practice for preparation and use of masonry mortars*. Bureau of Indian Standards.
20. Hamid, A. A., Abboud, B. E., & Harris, H. G. (1985). direct modeling of concrete block masonry under axial compression. In J. C. Grogan & J. T. Conway (Eds.), *Masonry: research, application, and problems, ASTM STP 87* (pp. 151–166). Philadelphia: American Society for Testing and Materials.
21. MSJC. (2002). *Building code requirements for masonry structures-ACI 530-02/ASCE 5-02/TMS 402-02*. Masonry Standards Joint Committee.

Numerical Investigation of Punching Shear Strengthening Techniques for Flat Slabs

Navya S. Ravi and Milu Mary Jacob

Abstract Flat slab structures situated in seismic zones, the moments transferring from slab to column through shear increases furthermore and becoming more tendency to punching shear failure during earthquakes. There are many ways to increase the punching shear strength of concrete slabs. Few techniques are shear reinforcement system, flat slab with externally bonded CFRP at corner columns, using post-installed steel bolts using carbon fiber reinforced polymer laminates in slabs, using glass fiber reinforced polymer laminates in slabs. A nonlinear static analysis of flat slab is done using ANSYS workbench to study punching shear strengthening techniques. Deflection values of various strengthening techniques were found out and compared with the limit. In this project, it is found that strengthening using carbon fiber reinforced polymer laminates and shear reinforcement are effective punching. Shear strengthening techniques.

Keywords Carbon fiber reinforced polymer laminates · Glass fiber reinforced polymer laminates · Shear reinforcement · Steel bolt · Punching shear strength

1 Introduction

There are many ways to increase the punching shear strength of concrete slabs. Few techniques are shear-reinforcement system, slab at. Corner columns with externally bonded CFRP, using post-installed steel bolts, carbon fiber reinforced polymer laminates in slabs, and glass fiber reinforced polymer laminates in slabs. Numerical analysis is done using finite element method (ANSYS) to study punching shear strengthening techniques.

N. S. Ravi (✉) · M. M. Jacob
Department of Civil Engineering, Saintgits College of Engineering, Ktu University, Kottayam, Kerala, India
e-mail: navya.sr1820@saintgits.org

M. M. Jacob
e-mail: milu.mary@saintgits.org

Table 1 Total deformation and equivalent elastic strain

Properties	Details
Material properties	M40 grade concrete
	Tensile strength (MPa) −2.9
	Modulus of elasticity (GPa) −28.3
Bottom steel reinforcement	6 mm @ 90 mm with concrete clear cover of 10-mm
Top steel reinforcement	4 of 6 mm @ 200 mm over each column extending 650 mm in each direction with concrete cover of 10- mm
Column reinforcement and stirrups	Longitudinal reinforcement: 4 of 12-mm stirrups: 6 mm distributed every 150-mm

2 Numerical Investigation

All the slabs were 2 m × 2 m square with a thickness of 125 mm. These slabs were attached to four 160 mm × 160 mm square, 720 mm high columns cast monolithically with the slab.

2.1 Properties Test Specimen

Properties of test specimen are shown in Table 1.

2.2 Modeling

The finite element software ANSYS was used in this study to simulate the structural behavior of the flat slabs. Because of the symmetry in loading and boundary conditions, a quarter of each slab specimen was modeled. All the four columns of each slab were designed to be pin supported at the base. Numerical analysis is done to study the following punching shear strengthening techniques for flat slab without column drop.

2.2.1 Slab Without Strengthening Methods

Slabs were 2 m × 2 m square with a thickness of 125 mm. These slabs were attached to four 160 × 160 mm square, 720 mm high columns cast monolithically with the slab. Because of the symmetry in loading and boundary conditions, a quarter of each slab specimen was modeled, and the slab model is shown in Fig. 1.

Loading was applied gradually at a rate of 3 kN/min. Apply a uniformly distributed load.

Fig. 1 Slab model

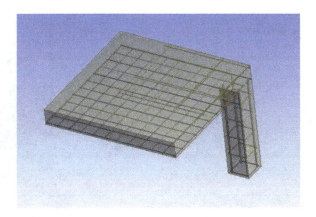

Fig. 2 Slab strengthened with CFRP sheets

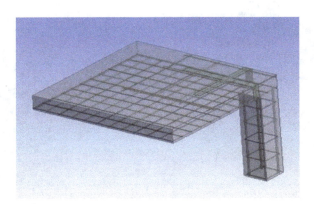

2.2.2 Flat Slab Strengthened with CFRP Sheets

Slab strengthened with carbon fiber polymer sheets of a width of 50 mm and thickness of 0.8 mm around the corner columns. The CFRP sheets were along the slab surface by between 340 and 500 mm for all strengthened slabs [1], and the slab model is shown in Fig. 2.

2.2.3 Strengthening Using Glass Fiber Reinforced Polymer Laminates in Slabs

Slab strengthened with glass fiber polymer sheets of a width 50 mm and thickness of 1.2 mm around the corner columns. The 3 GFRP sheets were the slab [2], and the slab model is shown in Fig. 3.

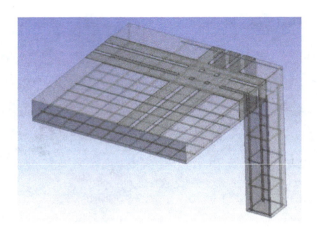

Fig. 3 Slab strengthening using GFRP

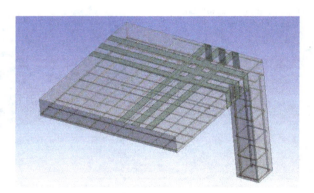

Fig. 4 Carbon fiber reinforced polymer laminates in slabs

2.2.4 Strengthening Using Carbon Fiber Reinforced Polymer Laminates

Reinforced polymer (FRP) flexible sheets on the slab around the column in two schemes with three layers of FRP/sheets glued to the tension face or both tension and compression faces of the slab [3], and the slab model is shown in Fig. 4.

2.2.5 Strengthening Using Post-installed Steel Bolts

Strengthening purposes, steel bolts were cut from M16 threaded bar grade 8.8. The yield and ultimate stress of the bolt were 640 and 800 MPa [4], and the slab model is shown in Fig. 5.

Fig. 5 Slab strengthening using post-installed steel bolts

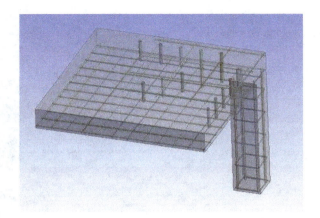

2.2.6 Shear Reinforcement and Shear Reinforcement with CFRP Laminates

Slabs were 2 m × 2 m square with a thickness of 125 mm. These slabs were attached to four 160 × 160 mm square, 720 mm high columns cast monolithically with the slab. Because of the symmetry in loading and boundary conditions, a quarter of each slab was modeled. Shear reinforcement provides with 12 mm diameter bars [5], and the slab model is shown in Fig. 7. Slabs were 2 m × 2 m square with a thickness of 125 mm. These slabs were attached to four 0.160 × 160 mm square, 720 mm high columns cast monolithically with the slab. Because of the symmetry in loading and boundary conditions, a quarter of each slab was modeled. Shear reinforcement provides with 12 mm diameter bars and CFRP laminates provided in three layers around column [4], and the slab model is shown in Figs. 6 and 7.

Fig. 6 Slab with shear reinforcement

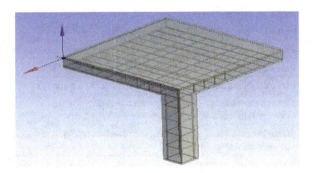

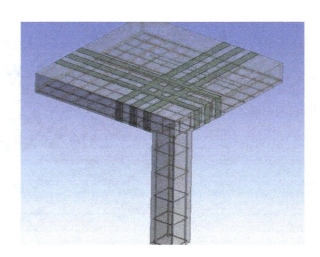

Fig. 7 Slab with shear reinforcement and CFRP

3 Results and Discussions

Numerical analysis of flat slab is done in ANSYS software. Total deformation and equivalent elastic strain were obtained. Total deformation and equivalent elastic strain of various strengthening methods are shown in Table 2. Load deflection curve of various punching shear strengthening methods is shown in Fig. 8.

Deflection control is an important serviceability consideration in the structural design of concrete buildings. The principal material parameters that influence concrete deflection are modulus of elasticity, modulus of rupture, creep, and shrinkage. After analysis, total deformation in slab without CFRP strengthening is 15.5618 mm. Based on IS 456: 2000 clause, 23.2 deflection should not normally exceed span/250.

Table 2 Total deformation and equivalent elastic strain

Flat slab	Total deformation (mm)	Equivalent elastic strain
Flat slab without strengthening methods	15.5618	0.0034
Flat slab strengthened with CFRP sheets	10.5204	0.0033
Strengthening using carbon fiber reinforced polymer laminates in slabs	7.4753	0.0033
Strengthening using glass fiber reinforced polymer laminates in slabs	7.5874	0.0034
Strengthening using post-installed steel bolts	7.6982	0.0030
Shear reinforcement in slab	7.5461	0.0031
Shear reinforcement with CFRP laminates	7.914	0.0039

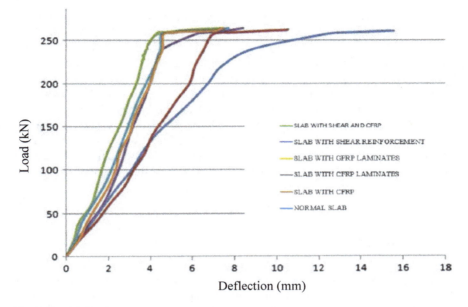

Fig. 8 Load deflection curve

Obtained deflection of flat slab without strengthening methods is higher than the limit. Equivalent strain is a scalar quantity also called the von Mises equivalent strain and is often used to describe the state of strain in concrete. After analysis, the equivalent elastic strain in concrete is 0.0034. Based on IS 456: 2000, equivalent elastic strain in concrete is 0.0035 in beam. Obtained value is within the limit.

Total deformation in slab with CFRP strengthening is 10.5204 mm. Strengthening corner slabs by surface mounted CFRP strips resulted in delaying the initiation of flexural cracks in slabs. This delay has improved the slab performance by total ultimate loading capacity of slab 260 kN and the sustained deflection from 15.5618 to 10.5204 mm. Strengthening using CFRP laminates reduce the deflection to desired limit. Equivalent elastic strain in concrete is 0.0033. Obtained value is within the limit.

Total deformation in slab with CFRP laminates is 7.4753 mm. Strengthening corner slabs by surface mounted CFRP strips resulted in delaying the initiation of flexural cracks in slabs. CFRP strips resulted in delaying the initiation of flexural cracks in slabs. This delay has improved the slab performance by total ultimate loading capacity of slab 260 kN and the sustained deflection from 15.585 to 7.4753 mm. Strengthening using CFRP laminates reduce the deflection to desired limit. Equivalent elastic strain in concrete is 0.0033. Obtained value is within the limit.

Total deformation in slab with GFRP strengthening is 7.5874 mm. Strengthening corner slabs by GFRP strips resulted in delaying the initiation of flexural cracks in slabs. Total deformation in using post-installed steel bolts is 7.6982 mm. Strengthening using post-installed steel bolts reduce the deflection to desired limit. Total deformation in using shear reinforcement is 7.5461 mm. Strengthening shear reinforcement reduces the deflection to desired limit. Total deformation in slab with CFRP and shear reinforcement is 7.1914 mm. Strengthening corner slabs by CFRP strips resulted in delaying the initiation of flexural cracks in slabs. This delay has improved the slab performance by total ultimate loading capacity of slab 260 kN and the sustained deflection from 15.585 to 7.1914 mm. Strengthening using stirrup and CFRP laminates reduce the deflection to desired limit. Figure 8 shows the load deflection curve of various strengthening methods. Graph also shows that the strengthening using stirrup and CFRP laminates have better punching shear strength than other methods.

4 Conclusions

Numerical analysis is done in ANSYS software to study the punching shear strengthening techniques for flat slab without column drop. Flat slab with various strengthening methods are analyzed, and mid-span deflection is varied for different strengthening techniques. Strengthening using carbon fiber reinforced polymer laminates in slabs reduce deflection to 32.39%. Shear reinforcement reduces deflection to 52.15%. When shear reinforcement provided with CFRP laminates, deflection is reduced to 53.77%. Strengthening using externally bonded CFRP reduce deflection to 51.96%. Glass fiber reinforced polymer laminates in slabs reduces deflection to 45.39%. Strengthening using steel bolts reduce deflection to 50.53%. Based on IS 456: 2000, clause 23.2 should not normally exceed span/250. Strengthening using CFRP laminates, post-installed steel bolts, shear reinforcement, and shear reinforcement with CFRP laminates reduce the deflection to desired limit. Strengthening using carbon fiber reinforced polymer laminates and shear reinforcement effective punching shear strengthening technique.

Acknowledgements We extremely thankful to everyone who gave their valuable guidance, encouragement, and immense support throughout our humble endeavor.

References

1. Abdulrahman, B. (2017). Experimental and numerical investigation into strengthening flat slabs at corner columns with externally bonded CFRP. *Construction and Building Materials, 139*, 132–147.
2. Salakawy, E., & Soudki, K. (2004). Punching shear behavior of flat slabs strengthened with fiber reinforced polymer laminates. *Journal of Composites for Construction ASCE*, ISSN 1090-0268/2004/5-384

3. Hassan, N., et al. (2018). Enhancement of punching shear strength of flat slabs using shear-band reinforcement. *HBRC Journal, 14,* 393–399.
4. Saleh, H., et.al. (2018). Experimental and numerical study into the punching shear strengthening of RC flat slabs using post-installed steel bolts. *Construct. Building Materials, 188,* 28–39
5. Almeida, A. (2016). Punching behaviour of RC flat slabs under reversed horizontal cyclic loading. *Engineering Structures, 117*(2016), 204–219.

Investigation of Bolted Beam–Column Steel Connections with RBS Subjected to Cyclic Loading

Deepa P. Antoo and Asha Joseph

Abstract Widespread damage to steel moment structures during the 1994 Northridge earthquake led researchers to develop alternative designs to the prescriptive pre-Northridge moment connection. Two key concepts have developed to provide a highly ductile response and reliable performance, strengthening the connection and weakening the beam, in order to avoid damages of column. The reduced beam section (RBS) allows controlled yielding of the beam by moving the plastic hinge region at the beam—in a short distance from the column's face—protecting the connection from any type of failure. Various shape cutouts are possible to reduce the cross-sectional area. The extended end plate bolted connection was chosen as an important type of RBS. In this paper, dynamic performance investigation of bolted beam–column steel connections with RBS was done by means of a three-dimensional FE model using ANSYS 16.1 Workbench. The performance of bolted beam–column connection with different RBS cut under cyclic loading was investigated.

Keywords Bolted beam–column connection · Reduced beam section · Finite element model · Cyclic loading

1 Introduction

As a response to the widespread damage in connections of steel moment-resisting frames that occurred during the 1994 Northridge, California and the 1995 Kobe, Japan earthquakes, a number of improved beam-to-column connection design strategies have been proposed [1]. Extended end-plate moment connections can be designed to be suitable for use in seismic force-resisting moment frames [2]. Connections should be designed to be stronger than the connecting beam to ensure that beam flange

D. P. Antoo (✉) · A. Joseph
Federal Institute of Science and Technology, Angamaly, Ernakulam, India
e-mail: deepapantoo1996@gmail.com

A. Joseph
e-mail: ashameledath@gmail.com

and web local buckling, a ductile and reliable limit state, will limit the strength of the beam-to-column connection assembly creating the "weak beam–strong column" mechanism [3].

Reduced beam section with bolted web (RBS-B) moment connections was one of the newly developed connections after the Northridge earthquake. The weakening of specific sections of the beam in order to change them into reliable energy dissipative zones, in case of an earthquake, is an idea that was developed in principle in the 1980s by Andre Plumier. The reduced beam section (RBS) allows controlled yielding of the beam by moving the plastic hinge region at the beam—in a short distance from the column's face—protecting the connection from any type of failure [1].

Even after the formation of hinges in the RBS zone, the connection remains in the elastic zone; hence, local failure is presented, and failed local member can be replaced [1]. Various shape cutouts are possible (constant, tapered or radius cut) to reduce the cross-sectional area. The bolted extended end plate connection was chosen as an important type of RBS connection due to extensive use in the European section profiles. RBS moment-resisting connection improves the ductility of steel member [4]. Improved ductility characteristics and cost effectiveness makes RBS preferable in seismic region [5]. Typical geometry for the design of radius cut RBS is shown in Fig. 1. Geometrical details of RBS are given in Table 1. This paper is concerned with RBS performance in bolted beam–column connection with different RBS cut.

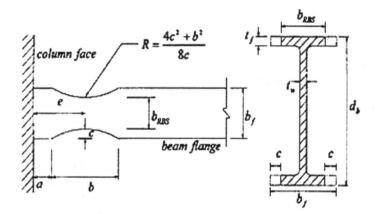

Fig. 1 Typical geometry of the radius cut RBS [6]

Table 1 Geometrical details of RBS

Parameters	Geometry details
a	Distance of the beginning of the RBS from the column face
b	Length of the RBS
b_f	Flange width
d_b	Beam depth
C	Depth of the flange cut
R	Radius of the cuts in both top and bottom flanges at the RBS
S	Distance of the intended plastic hinge at the center of the RBS from the column face, $s = a + b/2$

2 Modeling of Specimens

2.1 Introduction

Steel structures are very commonly used all over in the world due to its short construction time period and high strength. The connections provided in steel structures play an important role in its strength. So, the analysis and design of the steel member connection arrangement are very important. Different methods can be used to study the behavior of beam-to-column joints [7]. Finite element analysis as used in structural engineering determines the overall behavior of a structure by dividing it into a number of simple elements, each of which has well-defined mechanical and physical properties. Modeling is a very important aspect in finite element analysis; accuracy of the results depends on the accuracy in modeling. The accuracy of results is also dependent on the type and size of meshing used in the model. The finite element modeling and analysis of present study is carried out using ANSYS 16.1 Workbench.

A beam-to-column bolted connection is modeled with an extended endplate, where the cross-sections were HE 300 B and HE 160 A for the column and the beam, respectively. Double web plates of 12 mm thickness and continuity stiffeners equal to the beam flange thickness were assembled to the column strengthening furthermore the panel zone. Web plates of $400 \times 208 \times 12$mm, continuity plates $262 \times 100 \times 10$mm, extended end plates $250 \times 310 \times 20$mm, and 8 mm end plate welds were provided. Also M20, 10.9 grade bolts were used in the applied connection.

2.2 Loading and Boundary Conditions

Boundary conditions of the finite element model are shown in Fig. 2. The top and bottom of the column are provided with fixed boundary conditions thereby restricting the translations and rotations in x-, y-, and z-directions. For determining the seismic behavior of different configurations in RBS, cyclic loading was applied at the beam

Fig. 2 Boundary conditions

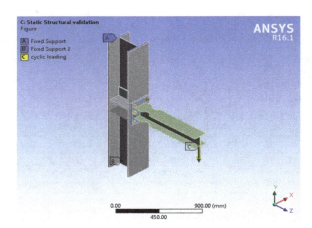

Table 2 AISC 2002 loading schedule

Load step	Peak deformation, δ_y	Number of cycles
1	±0.375	6
2	±0.50	2
3	±0.75	2
4	±1.00	4
5	±1.50	2
6	±2.00	2
7	±3.00	2
8	±4.00	2

tip. This loading protocol was developed to obtain the behavior of beam–column moment connections. The details of loading sequence for AISC loading protocol are presented in Table 2.

2.3 Connections and Contacts in the Model

Different types of contacts are provided between the connections. Frictionless contact is provided between columns to extended end plate connections. Frictional contact is provided between extended end plate-to-bolt, column-to-bolt connections, and also in the column-to-nut connections. Similarly, bonded contacts are provided between column-to-web plates, column-to-continuity plates, bolt-to-nut, web plate-to-continuity plate, and also in extended end plate-to-beam connections.

2.4 Geometric Modeling

The study aims to develop a better steel moment resisting connection with the reduced beam sections. Different configurations are tested to evaluate the behavior (Figs. 3, 4, 5 and 6). RBS with circular, trapezoidal, rectangular cut outs are made near the column face in the flanges of the beam. Area of the reduced section for all configurations is made constant. The modeling includes the modeling of beam, column, and reduced sections of beam. The geometry is defined for each configuration, and suitable properties are assigned to them which is as given in Table 3.

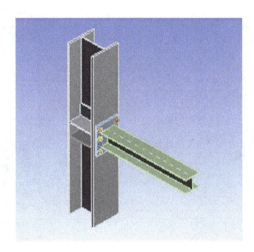

Fig. 3 FE model without RBS

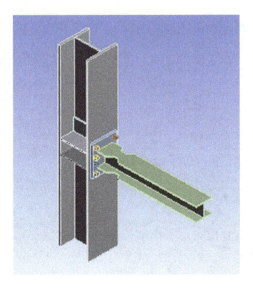

Fig. 4 RBS with trapezoidal cut

Fig. 5 RBS with circular cut

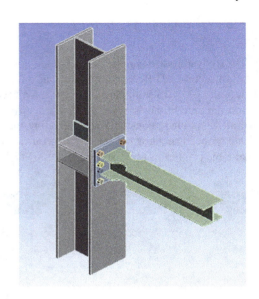

Fig. 6 RBS with rectangular cut

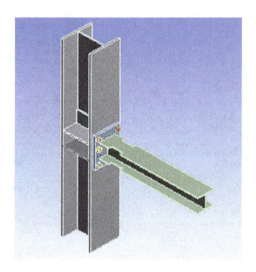

3 Results and Discussions

3.1 General

The behavior of different configurations of connection has been studied. The parameters such as stress distribution, load-carrying capacity, deformation, and energy

Table 3 Dimensions of RBS with variable cut outs

	RBS with circular cut	RBS with rectangular cut	RBS with trapezoidal cut
Distance of the beginning of the RBS from column face, a (mm)	96	96	96
Area of RBS (mm^2)	3321.7	3321.7	3321.7
Length of the RBS, b (mm)	114	114	$b_1 = 114$ $b_2 = 52.08$
Depth of the flange cut, c (mm)	40	29.137	40
Radius of the cuts in both top and bottom flanges at the RBS, r (mm)	60.61	–	–

absorption are studied. The location of plastic hinge is identified in each configuration. According to the capacity design concept, the formation of plastic hinge should be away from the column face [8]. The seismic performance of all the configurations is studied by applying a cyclic loading at the tip end of the beam. AISC loading protocol was used for this purpose. Load versus deformation curve is plotted in order to obtain the energy absorption.

3.2 Stress Distribution

The structure is analyzed after the application of the cyclic loading at the free end of the beam. The stress distribution for each configuration is noted. According to the capacity design concept, the location of the plastic hinge should be away from the column face. The maximum stress in column and beam in each configuration is compared.

3.2.1 Beam–Column Joint Without RBS

By performing the cyclic analysis, it is found that the major stress is concentrated at the column face, which is not recommended for the seismic performance. The stress distribution for beam–column joint without RBS is shown in Fig. 7.

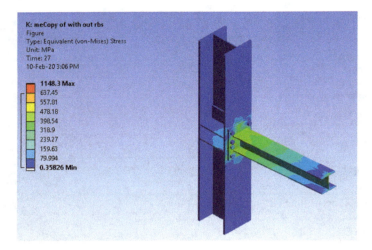

Fig. 7 Stress distribution of beam–column joint without RBS

3.2.2 RBS with Trapezoidal Cut

Upon cyclic loading, maximum stress in beam and column is 555.61 MPa and 461.36 MPa, respectively. Stress of 874.55 MPa is found to occur in bolt. The location of hinge is changed from the column face to the RBS regions, which it is good for the seismic performance. The stress distribution for the reduced beam section with trapezoidal cut is shown in Fig. 8.

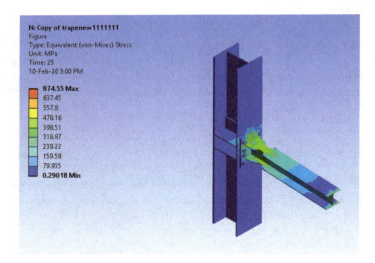

Fig. 8 Stress distribution of RBS with trapezoidal cut

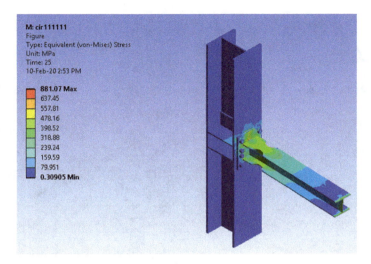

Fig. 9 Stress distribution of RBS with circular cut

3.2.3 RBS with Circular Cut

Under the cyclic loading, maximum stress of 555.36 MPa is occurred at the reduced beam section and 480.46 MPa at the column. Stress of 881.07 MPa is found to occur in the bolt. For this configuration also, the hinge position is changed from the column face to the RBS region [9]. The stress distribution for the reduced beam section with circular cut is shown in Fig. 9.

3.2.4 RBS with Rectangular Cut

For RBS with rectangular cut, the maximum stress in beam is 540.06 MPa and in column is 514.85 MPa. The stress in column is much higher than maximum column stress with circular cut. The maximum stress of 888.37 MPa occurs in the bolt. The stress distribution for the reduced beam section with rectangular cut is shown in Fig. 10.

3.3 *Energy Absorption Capacity*

Load v/s deformation is plotted to study the seismic behavior of the selected configuration. From the hysteresis graph, we can determine the energy absorption capacity of each configuration [10]. Area under the curve gives the energy absorption values. The curves obtained are shown in Fig. 11, 12, 13 and 14. The energy absorption corresponding to each model is given in Table 4.

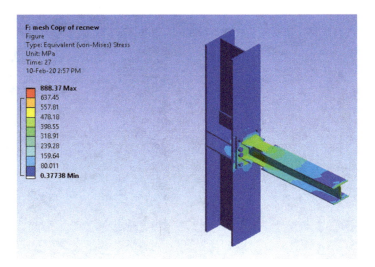

Fig. 10 Stress distribution of RBS with rectangular cut

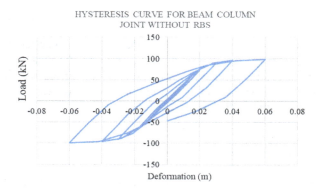

Fig. 11 Hysteresis curve for beam–column joint without RBS

4 Interpretation of Analysis Results

The response parameters to cyclic loading such as deformation, stress, load-carrying capacity, and energy absorption corresponding to each model are tabulated in Table 4.

The energy absorption of RBS with circular cut is 18.901 kJ, which is 35.24% higher than conventional connection. Also by providing circular cut, the stress in column is reduced by 7.9%, and plastic hinge location is shifted to reduced beam section. Hence, it is recommended that RBS with circular cut gives better performance.

Investigation of Bolted Beam ... 143

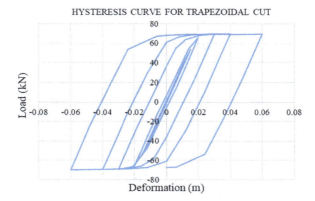

Fig. 12 Hysteresis curve for RBS with trapezoidal cut

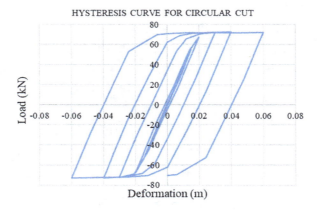

Fig. 13 Hysteresis curve for RBS with circular cut

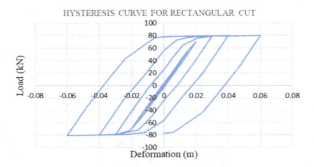

Fig. 14 Hysteresis curve for RBS with rectangular cut

Table 4 Interpretation of analysis results

Cross section	Yield load (kN)	Ultimate load (kN)	Maximum		Yield deformation (mm)	Ultimate deformation (mm)	Energy absorption (kJ)
			Column stress (MPa)	Beam stress (MPa)			
Without RBS	30.657	99.864	521.71	460.93	7.5	60	13.975
Trapezoidal	27.802	69.833	461.36	555.61	7.5	60	18.80
Circular	28.102	72.667	480.46	555.36	7.5	60	18.901
Rectangular	28.249	79.985	514.85	540.06	7.5	60	18.322

5 Conclusions

It is essential for the connection to be sufficiently strong and robust, mobilizing the stresses to a desired location along the length of the beam away from the connection assembly, creating the "weak beam–strong column" mechanism.

In this paper, the different RBS configurations and the behavior of RBS moment connection with extended bolted end plate that is widely used in beam-to-column moment connections subjected to cyclic loading were examined, and following conclusions were made:

- By adopting reduced beam section, the plastic hinge location is shifted from the column face to the RBS region.
- Considering the stress concentration of the column and energy absorption, the circular configuration is found to be better model.
- The column stress in circular configuration was reduced by 7.9% on comparing with the conventional moment connection, and it has a load-carrying capacity of 72.667 kN.
- Increase in energy absorption by 35.24% was observed in circular cut when compared with conventional connection.

References

1. Sofias, C. E., Kalfas, C. N., & Pachoumis, D. T. (2014). Experimental and FEM analysis of reduced beam section moment endplate connections under cyclic loading. *Engineering Structures, 59*, 320–329.
2. Abidelah, A., Bouchaïr, A., & Kerdal, D. E. (2012). Experimental and analytical behavior of bolted end-plate connections with or without stiffeners. *Journal of Constructional Steel Research, 76*, 13–27.
3. Sumner, E. A., & Murray, T. M. (2002). Behavior of extended end-plate moment connections subject to cyclic loading. *Journal of Structural Engineering, 128*(4), 501–508.
4. Swati, A. K., & Gaurang, V. (2014). Study of steel moment connection with and without reduced beam section. *Case Studies in Structural Engineering, 1*, 26–31.
5. Han, S. W., Moon, K.-H., Hwang, S.-H., & Stojadinovic, B. (2012). Rotation capacities of reduced beam section with bolted web (RBS-B) connections. *Journal of Constructional Steel Research, 70*, 256–263.
6. Christos, S., & Tzourmakliotou, D. (2018). Reduced beam Section (RBS) moment connections-analytical investigation using finite element method. *Civil Engineering Journal, 4*(6), 1240–1253.
7. Kulkarni, R. B., & Vaghe, V. M. (2014). Experimental study of bolted connections using light gauge channel sections and packing plates at the joints. *International Journal of Advanced Structural Engineering (IJASE), 6*(4), 105–119.
8. Lee, C.-H., Jeon, S.-W., Kim, J.-H., & Uang, C.-M. (2005). Effects of panel zone strength and beam web connection method on seismic performance of reduced beam section steel moment connections. *Journal of Structural Engineering, 131*(12), 1854–1865.

9. Sophianopoulos, D. S., & Deri, A. E. (2019). Steel beam-to-column RBS connections: FEM analysis under cyclic loading. *World Journal of Mechanics, 9,* 17–28.
10. Kim, J., Ghaboussi, J., & Elnashai, Amr S. (2012). Hysteretic mechanical–informational modeling of bolted steel frame connections. *Engineering Structures, 45,* 1–11.

Effect of Shock Absorbers in Enhancing the Earthquake Resistance of a Multi-storeyed Framed Building

Deepa Balakrishnan, Anjali, and Salauddin

Abstract The concept of base isolation systems has gained wide spread acceptance in enhancing the earthquake resistance of a structure. To accomplish the predicted behaviour of the base-isolated buildings, the design for base isolation system is regarded as the dominant factor. The base isolation of the structure basically reduces the storey shear and acceleration and increases time period, storey displacement and storey drift inducing flexibility in rigid structure by dissipating energy to foundation. For the study, three framed structures in zone-V with vertical and horizontal irregularity having G+10 storey have been analysed for its seismic behaviour with and without isolator using ANSYS. The analysis incorporating earthquake load is based on the seismic coefficient method as described in IS 1893: 2002. The study has been conducted in complete conformity with various provisions in Indian Standards as Code of Practice for plain and reinforced concrete IS 456-2000 [1].

Keywords Base isolation · Lead–rubber bearing · Vulcanized rubber · Base shear · Storey displacement · Storey drift · Natural period · Free vibration analysis

1 Introduction

Base isolation technique is commonly adopted as safety precaution in earthquake prone areas all over the world. It is implemented in the foundation section of the structure to reduce the effects and damages caused by an earthquake. This system is designed to take the weight of the building and permits translation movements in the foundations during the earthquake. Seismic isolation can improve the seismic performance of the structure and also minimize damage [2]. The principle of seismic isolation is to introduce flexibility at the base of a structure in the horizontal plane, while at the same time, it introduce damping elements to restrict the amplitude of the motion caused by the earthquake. New impetus was given to the concept of seismic isolation by the successful development of mechanical energy dissipaters

D. Balakrishnan (✉) · Anjali · Salauddin
Cochin University of Science and Technology, Kochi, Kerala 682022, India
e-mail: deepa_balu@cusat.ac.in

and elastomers with high damping properties. Mechanical energy dissipaters, when used in combination with a flexible isolation device, can control the response of the structure by limiting displacements and forces, thereby significantly improving seismic performance. The seismic energy is dissipated in components specifically designed for that purpose, relieving structural elements, such as beams and columns, from energy-dissipation roles. By using rubber elastomer for base isolation, it is possible to avoid large plastic deformation of moment resisting frame and reduce shear resulting from large scale earthquake [3].

Conventional construction can cause very high floor accelerations in stiff buildings and large inter storey drifts in flexible structures. These two factors cause difficulties in ensuring the safety of the building components and contents [4]. The base-isolated building retains its original shape while the lead–rubber bearings supporting the building are deformed. The deformation of base-isolated building itself escapes damage—which implies that the inertial forces acting on the base-isolated building have been reduced. The analysis and observations of base-isolated buildings in earthquakes have been shown to reduce fixed-base buildings displacement to comparable limits [5]. Seismic isolation achieves a reduction in earthquake forces by lengthening the period of vibration at which the structure responds to the earthquake motions. The most significant benefits obtained from isolation are in structures for which the fundamental period of vibration without base isolation is short, i.e. less than 1 s. The natural period of a building generally increases with increasing height. Taller buildings reach a limit at which the natural period is long enough to attract low earthquake forces without isolation. Therefore, seismic isolation is most applicable to low-rise and medium-rise buildings and become less effective for high-rise ones [3].

2 Experimental Investigations

2.1 Methodology of Work

STEP 1: Selection of the Building

Three different structural configurations were selected as the seismic response of building depends upon its plan and elevation configuration. All the structures were residential type with G + 10 floors in total located in seismic zone-V on medium type soil. First structure is cuboidal with rectangular plan and regular elevation. Second structure is irregular in plan having plus shape and is regular in elevation. Third structure is L-shaped while considering elevation.

Fig. 1 Details of rubber bearing

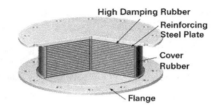

STEP 2: Load Calculation

Dead and live load on the structure were calculated as per Code of Practice for Design Loads IS 875 (Part 1)-1987, IS 875 (Part 2)-1987. Earthquake load calculation is based on seismic coefficient design method as described in IS 1893: 2002 [6].

STEP 3: Design of Isolator

Lead–rubber bearing is used as a base isolation in which the bearings are stiff in vertical direction and very flexible in horizontal direction. The vertical stiffness is achieved through the steel plate layers of 5 mm, which is laid between each alternative layer; it does not let the rubber to displace from their position by holding them tight. Lead is chosen because of its plastic property by which it has the capability of deforming repeatedly without losing its strength. This is advantageous during seismic activity since it will return back to its original state after deformation. Vulcanized rubber is used because of its excellent elasticity, low water absorption tendency and high durability as compared to natural rubber [4]. The details of bearing used are shown in Fig. 1.

Preliminary design of lead–rubber isolator was done on the basis of natural period and frequency of structure and the thickness of isolator was fixed after many iterations. Final properties and dimension of lead–rubber isolator adopted for all the three structures are:

- Modulus of elasticity = 54.93 Gpa
- Density = 3440.96 kg/m^3
- Poisons ratio = 0.439
- Lead core diameter = 10 cms
- Diameter of isolator = 30 cms
- Total diameter with 1.5 cms outer cover = 33 cms
- Thickness of rubber = 30 cm (60 layers of 5 mm thickness)
- Thickness of steel shim = 17.7 cm (59 layers of 3 mm thickness)
- Total thickness of isolator = 47.7 cm.

STEP 4: Modelling and Analysis of Building:

Structural analysis software ANSYS has been used for the modelling and analysis of the building with and without base isolation systems.

3 Results and Discussion

All the three structures were modelled in ANSYS with and without isolators and loading conditions as per clause 6.3.1.2 as given in IS 1893 (part 1): 2002 [6] were applied. Isolators were provided at two different positions—in basement column and at bottom of ground floor column successively. The observed results are discussed in the subsequent sections.

1. *Variation in natural period of structures*:
   The vibration of a building consists of a fundamental mode of vibration and the additional contribution of various modes, which vibrates at higher frequencies. Time period corresponding to fundamental mode of vibration is taken as natural period of building. Free vibration analysis was done to find natural frequency and time period of structure.
   When isolators were incorporated in the structure, natural time period increased to more than 1.5 s in each case. Thus, the seismic isolation achieves a reduction in earthquake forces by lengthening the period of vibration at which the structure responds to the earthquake motions. Since the fundamental period of vibration without base isolation was less than 1 s in all structures significant benefits can be obtained from isolation. Isolators provided in ground floor column are more effective in increasing the time period. The results are given in Table 1.
2. *Reduction in base shear*:
   When isolators were incorporated in the structure, the base shear value reduced tremendously. Base shear reduced more in the case when isolators were located in basement column. In all the three structures, base shear reduced by approximately 95% when isolators were provided in basement column is given in Table 2.
3. *Variation in storey displacement and storey drift*:
   Figures 2, 3 and 4 show the variation in storey displacement and storey drift for structures 1, 2 and 3 with and without shock absorbers. It has been observed

Table 1 Natural period of vibrations for different structural configurations

Type of structure	Natural period of vibrations (s)			% increase in natural period of vibration	
	Without isolator	With isolator (at basement column)	With isolator (at GF column)	With isolator (at basement column)	With isolator (at GF column)
Structure 1 (Regular)	0.96	1.757	1.88	83.02	95.83
Structure 2 (Plan irregularity)	0.958	1.49	1.88	55.5	96.24
Structure 3 (Vertical irregularity)	0.85	1.596	1.72	87.76	102.35

Table 2 Base shear for different structural configurations

Type of structure	Base shear (kN)			% reduction in base shear	
	Without isolator	With isolator (at basement column)	With isolator (at GF column)	With isolator (at basement column)	With isolator (at GF column)
Structure 1 (Regular)	6205	270	910	95.65	85.33
Structure 2 (Plan irregularity)	5050	176	682	96.5	86.5
Structure 3 (Vertical irregularity)	5010	223	590	95.5	88.2

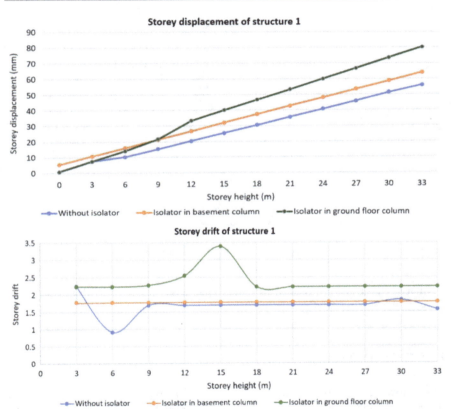

Fig. 2 Variation in storey displacement and storey drift for structure 1

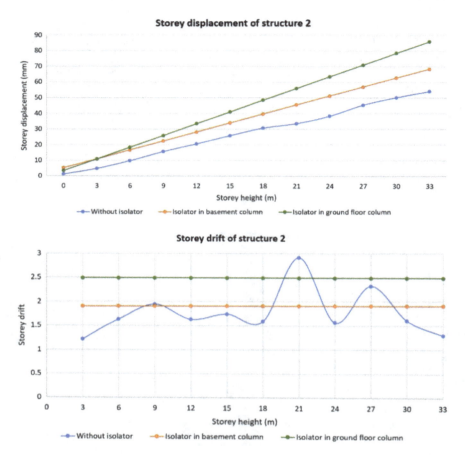

Fig. 3 Variation in storey displacement and storey drift for structure 2

that in all the cases, storey displacement increased by 15–20% while storey drift increased by 40–60%.

4 Conclusions

The following conclusions have been drawn from the study of effect of shock absorbers in enhancing the earthquake resistance of a multi-storeyed framed building:

1. Regular structure performs well in earthquake as compared to a structure with irregularity (a building that lacks symmetry and has discontinuity in geometry, mass or load resisting elements).
2. Maximum deflection occurred in structure 3 with irregular elevation and was found to be 1.11 times the minimum deflection that occurred in structure 2.

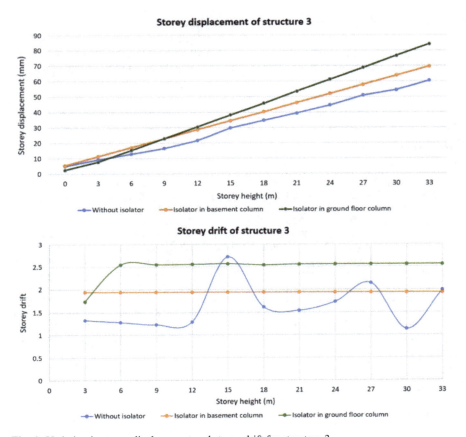

Fig. 4 Variation in storey displacement and storey drift for structure 3

Hence, it can be concluded that the elevation irregularity causes more damage to a structure as compared to plan irregularity.
3. Natural time period of building which was less than 1 s without base isolation increased by about 95% by isolating the base. Thus, the seismic isolation achieves a reduction in earthquake forces by lengthening the period of vibration at which the structure responds to the earthquake motions.
4. Due to lengthening of period of vibration, base shear value is reduced. Base shear reduced more in the case when isolators were located in basement column. In all the three cases, base shear reduced by around 95% when isolators were provided in basement column. While base shear reduced by around 85% when isolators were provided in ground floor column.

5. Base isolation increased storey displacement in all cases. Maximum storey displacement was seen in the case when isolators were in the ground floor columns. In structure 1, 2 and 3, storey displacement increased by 15, 26 and 16% when isolators were provided in basement column, whereas it increased by 43, 59 and 40% when isolators were provided in ground floor column.

References

1. IS-456: Indian standard for plain and reinforced concrete- code of practice.
2. Wiles, J. I., Stephens, S. F. & S. E. (2012). An overview of the technology and design of base isolated buildings in high seismic regions in the united states. *International Journal of Scientific and Engineering Research, 3*(3).
3. Wu, T.-C. (2001). 'Design of Base Isolation System for Buildings', Bachelor of Science in Civil Engineering Thesis, Chung-Yuan Christian University, Chung-Li, Taiwan.
4. Chowdhury, M. A., & Hassan, W. (2013). Comparative study of the dynamic analysis of multi-storey Irregular building with or without Base Isolator. *International Journal of Scientific Engineering and Technology, 2*(9).
5. Landge, M. S., & Joshi, P. K. (2017). Comparative study of various types of dampers used for multi-story R.C.C. building. *International Journal for Research in Applied Science & Engineering Technology (IJRASET), 5*(IV).
6. IS 1893 (Part 1): 2002 – Indian standard criteria for earthquake resistant design of structures, Part 1 General Provisions

Review Paper on Pavement Condition Assessment

Saranya Ullas and C. S. Bindu

Abstract India has the second-largest road network system in the world. Roads are the most reliable and accessible system for transportation which determines the growth of the country. After construction, its condition goes on deteriorating with time due to several factors (load factors, environmental factors, etc.). Pavement performance prediction models developed by the researchers include models related to rutting, cracking, potholes, etc., and are applicable only for a specific set of traffic or environmental conditions. Thus, these models lack universal acceptance and applications. Development of realistic and efficient pavement condition rating model is very essential to determine the future condition of the pavement and its service life. An effort has been taken to identify the various parameters that affect the condition of the pavement and different tools used for modeling based on the literature review. The different tools for modeling include regression models, Markovian probabilistic process, multi-attribute utility theory, soft computing techniques, etc.

Keywords Deterioration model · Service life · Parameters

1 Introduction

Pavement management system effectively manages all phases during the lifetime of the pavement. Its function includes assessing the condition of the pavement and to decide the maintenance and rehabilitation strategies for the pavement. Pavement condition assessment is a vital measure to quantify the existing condition of pavement and to predict the overall performance of the highway.

S. Ullas (✉) · C. S. Bindu
Department of Civil Engineering, School of Engineering, CUSAT, Kochi, India
e-mail: saranyaullas@gmail.com

C. S. Bindu
e-mail: binduromeo@gmail.com

According to American Association of State Highway and Transport Officials (AASHTO), pavement performance is defined as *"The serviceability trend of pavement over a design period of time, where serviceability indicates the ability of the pavement to serve the demand of the traffic in the existing conditions."* Numbers of pavement performance models were developed by the researchers to predict future pavement conditions and to select the best rehabilitation activities.

Thus, the development of an efficient and realistic model for the prediction of future condition is beneficial to highway maintenance agencies.

2 Pavement Performance Models

Pavement performance models can be considered as an equation or relation which connects intrinsic and extrinsic factors. Depending upon the factors selected for the development, models are categorized into (a) surface characteristic-based models, (b) models based on environmental factors, and (c) pavement performance rating models [1].

Some researchers have categorized it into two; deterministic and probabilistic models. Saha et al. developed deterministic techniques and are most common type of pavement deterioration model. It includes straight line extrapolation, S-shaped curves, polynomial-constrained least squares, and logistic growth model. But these models cannot take the uncertainties in pavement behavior under variable traffic load and weather conditions. Uncertainties require the use of probabilistic models for pavement deterioration. Markov model comes under probabilistic approach [2].

2.1 Surface Characteristics Based Models

Such models are developed based on the surface characteristics of the pavement. It includes models related to roughness, rut depth, raveling, potholes, etc., and is generated by considering the traffic and age as factors [1].

Chopra et al. [3] introduced five distress-based models. It includes models for (a) cracking progression, (b) raveling progression, (c) pothole progression, (d) rutting progression, and (e) roughness progression. The factors included in the models were modified structural number, age, present condition of the distress, and equivalent standard axle repetition in the analysis year.

Jeong et al. [4] developed distress-based model using fuzzy inference system which can consider the uncertainties and unavailability of information. Gupta et al. [5] formulated models for deflection and riding quality and developed a maintenance management plan for low volume roads.

2.2 Models Based on Environmental Characteristics

These models incorporate the effect of environmental factors such as moisture, temperature, freeze and thaw cycles, pavement layer and surroundings, precipitation, movement of groundwater, etc., on the performance of the pavements [1].

Abu samra et al. [6] developed a comprehensive condition rating model that incorporates climatic factors, physical properties, and operational factors. Based on climatic factors, two models were developed one for winter and another for summer. But, the major limitation was models that were only applicable to full depth asphalt pavement (with base and surface layer).

Probabilistic models developed by Saha et al. incorporated the uncertainties in the pavement deterioration model connected to traffic and weather. Researchers incorporated the uncertainties (traffic and weather) into account while modeling a probabilistic pavement performance model. The results showed an improved curve after the selection of uncertainties [2].

2.3 Pavement Performance Rating (PPR) Models

Such models define the pavement performance using weighted values. Weighted values have certain range, which will be defined by the researcher. Several indices are proposed by different researchers to determine the condition of the pavement. Pavement condition index (PCI), present serviceability index (PSI), international roughness index (IRI) and present serviceability rating (PSR), surface distress index (SDI), etc., come under this category [1].

Condition rating models were developed based on pavement type, thickness, traffic, pavement age, and current condition and its value varies from 1 to 9. Model was based on the type, amount, and severity of evident pavement distresses, as well as the overall roughness of the pavement surface, level of wheel path rutting, and magnitude of transverse joint faulting. Old condition rating survey (CRS) prediction models were focused on the age of the pavement only, and new CRS model could incorporate the factors such as traffic and pavement structure [7].

Pavement condition index developed by Suroyoto et al. derived the correlation between IRI, PCI, and SDI. Correlation was studied based on Pearson correlation analysis and found that a good correlation exists between PCI and IRI [8].

Pavement performance index (PPI) was developed [9] by considering the various deteriorating parameters from the detailed literature review. A simple formula (Eq. 1) was suggested by the researcher for PPI based on rating and weightage of each parameter

$$\text{PPI} = \sum R_i * W_i \tag{1}$$

PPI = Pavement performance index; R_i = rating of each deteriorating parameter, and W_i = weightage of each deteriorating parameter.

Shahnazuri et al. [10] developed an alternative approach for the development of pavement condition indices using artificial neural network (ANN) and genetic programming (GP). The study was based on PCI values of 12,487 road stretches, and these data sets were used for the development of the model. A computer program was developed using ANN and GP for calculating the PCI.

These models depict the functional performance of the pavements and will integrate the maintenance and rehabilitation strategies of the road stretches.

2.4 Factors Affecting the Condition of the Pavement

Detailed literature survey revealed that the identification of performance parameters is the most difficult task in pavement condition assessment. It depends upon the roadway factors, climatic factors, etc., and varies from place to place and time to time. Factors that affect the condition of the pavement include intrinsic and extrinsic factors in combination with performance indicators.

Researchers selected the major factors affecting the performance of pavement as traffic, moisture, subgrade, construction quality, and maintenance [11]. In addition to the above factors, Jeong et al. [4] selected climatic change as a factor of great importance during the development of impact assessment system. Abu samra et al. [6] modified own model based on climatic factors. Gupta et al. indicates for low volume roads age, and traffic is the most important performance indicator [5].

Park and Kim [12] classified the factors affecting the condition of the pavement as intrinsic and extrinsic. Intrinsic factors include material characteristics such as plastic deformation, fatigue, cracking, and thickness of the road base. Factors external to the pavement comes under extrinsic factors. Average traffic volume, rate of heavy vehicle traffic, and traffic speed comes under this category.

Singh et al. [13] considered international roughness index (IRI), surface modulus (Eo), rut depth (R), and friction coefficient (f) as the major parameters based on the functional and structural performance of the selected stretches pavement. Then, two soft computing techniques were applied to select the best maintenance and rehabilitation strategies.

Chopra et al. [3] developed model for modified structural number, age, present condition of the distress, and equivalent standard axle repetition in the analysis year.

Recent studies have revealed that the performance models not only depend upon the performance indicators but also affected by the characteristics of underlying layers. Luo et al. [14] developed a model based on material properties, material characteristics, and structural performance. An effort has been taken to identify the characteristics of underlying layers and how it is connected to the performance indicators (for both flexible and rigid pavements). Key factors for flexible pavements include the resilient modulus, shear strength, and permanent deformation. For rigid pavements, the resilient modulus, shear strength, erosion, and permanent deformation of unbound layers or subgrade are selected as critical factors.

3 Tools for Modeling

Different tools for pavement performance modeling include regression model, Markov chain model, multi-attribute utility theory, and soft computing techniques.

3.1 Regression Models

Regression models are statistical tools for deterioration modeling. It includes linear and nonlinear models. Formulation of regression model for the pavement performance is a difficult task since it involves a large number of independent variables and in some cases variables are not independent in nature.

Gupta et al. [5] formulated a regression model for the prediction of deflection and riding quality using regression techniques. Validation of these models was carried out using paired t-test and chi-squared test. As discussed earlier, researcher proposed a maintenance priority index using road condition index (RCI). RCI was calculated from deflection index and riding quality index.

Even though the quantification of pavement condition models by mathematical equations is a difficult task, a number of models have been available to predict the condition of the pavement. Nowadays, such models are not adopted for the deterioration modeling since it does not incorporate the uncertainties involved in modeling.

3.2 Markov Chain

Indices based system was introduced which incorporates transverse, longitudinal, fatigue cracking indices roughness, and rut indices. The indices usually vary from 0 to 100, where 100 represents a pavement in best condition and 0 represents a pavement in worst condition. Three curves were developed based on deterministic approach, but the authors were not able to incorporate the uncertainties connected to traffic and weather.

In Markov chain development, the first step is associated with the development of transition probability matrix (Fig. 1). It showed the probability that selected road stretches will remain in a specific condition in that specified year [2].

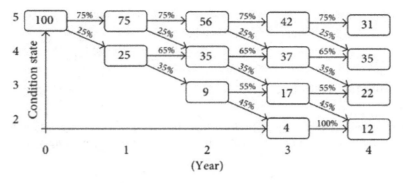

Fig. 1 Example of Markovian chain deterioration model

3.3 Multi-attribute Utility Theory (MAUT)

Multi-attribute utility theory (MUAT) uses utility functions, and it represents the preferences to be given to a set of decision attributes. Development of MAUT condition rating model includes factor weight and factor performance impact to calculate the condition of the pavement. Factor weight was computed from a survey conducted among 60 experts in the field. Analytic hierarchy process was used to assess the factor weight in MUAT model which represents the importance of one factor relative to the other factor [6].

3.4 Soft Computing

Soft computing techniques can deal with the uncertainties and approximations to solve complex problems involved in real life. It is based on techniques such as artificial neural networks, fuzzy logic, genetic algorithms, etc. [15].

3.4.1 Artificial Neural Network (ANN)

Artificial neural network models are nonlinear in structure, and it can form complex mathematical model. It consists of number of neurons as processing elements called nodes or units.

Shahnazuri et al. focused on the alternative approach to determine the future PCI values using techniques such as ANN and genetic programming (GP). Researchers studied the network based on feed-forward network and developed an ANN model and trained it by back propagation method. A four layer (an input layer, two hidden layers, and an output layer) network with log-sigmoid transfer function was used (Fig. 2). The output of this model was evaluated by mean square error [10].

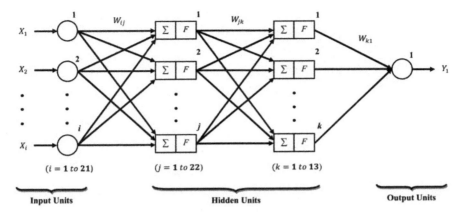

Fig. 2 Steps for artificial neural network model

ANN models were developed to determine the deflection and riding quality. The tool for training the data includes batch gradient descent with momentum algorithm. Models were developed using two different tools and suggested a maintenance management plan for low volume roads and expressed it as a maintenance priority index [5].

3.4.2 Fuzzy System

Nowadays, fuzzy is one of the most widely used techniques for the pavement condition assessment and prioritization. As discussed earlier, the factors affecting the pavement performance models are subjective in nature. In such cases, fuzzy logic proves to be a suitable tool for dealing with subjective judgment and uncertainty involved in the pavement condition assessment.

Jeong et al. proposed a fuzzy inference system as a method for developing climatic impact assessment system. System calculates the alternation in the rate of deterioration in relation with climate change. Fuzzy inference system explains the procedure in a four step process; (a) selection of the stretches and determination of baseline deterioration curve, (b) collection of expert knowledge regarding, (c) building of fuzzy inference system, and (d) evaluation of impacts of climatic change on the selected stretches [4].

Fuzzy analytical hierarchy process (FAHP) and fuzzy weighted approach FWA were the techniques used for the prioritization of the pavements. Analytic hierarchy process (AHP) is a decision-making tool which helps to take decisions in complex real-life problems. Incorporation of uncertainties is a difficult task in conventional AHP process; however, FAHP opens the way for including the vagueness and fuzziness into the model. Figure 3 shows the prioritization procedure adopted for the determination of the condition of the pavement [13].

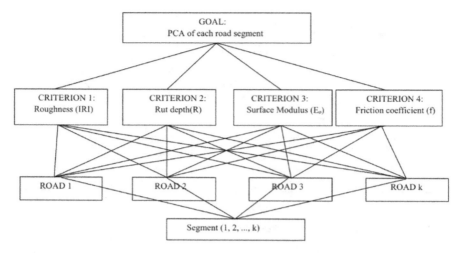

Fig. 3 Hierarchical architecture of criteria and alternatives

A model was formulated by a researcher for the determination of present and future performance of the pavement using fuzzy logic technique. The rule-based system developed for modeling yield a better result compared with another model of specific data set [16]. The inclusion and exclusion of performance parameter at any stage were the main advantage of this model.

3.4.3 Genetic Programming

Genetic programming models were developed [3], and the results obtained were represented in terms of coefficient of determination (R^2) and root mean square error (MS). Data collected during the period 2012–2015 on the selected stretches have been used for development and validation of the model. It was found that there exists a good relationship between the predicted and observed values.

4 Conclusions

Recent research papers in the area of pavement condition assessment were reviewed. Following are the conclusions:

- The pavement condition assessment models were classified into three and in surface characteristics model researchers developed models for distresses without considering the uncertainties involved. But this was the major gap in the development of surface characteristics based models.

- The above-discussed drawback was rectified in the models based on environmental characteristics. Such models showed better results compared with models of surface characteristics by the addition of uncertainties.
- PPR models are index-based systems and incorporate large number of performance parameters (both traffic and environmental). The major advantage of these models is that it can prioritize the maintenance and rehabilitation strategies of the selected road stretches and are widely adopted one.
- Apart from the conventional factors that affect the condition of the pavement, studies must be carried out to identify the other parameters. Recent studies have revealed that inclusion of factors such as climatic, physical, and operational factors is essential for better results.
- The tools for modeling depends upon the factors and subfactors selected, and soft computing techniques can handle this tremendous amount of uncertainties involved. It can handle n number of parameters, and addition and deletion of parameters are possible at any stage of modeling.

References

1. Gupta, A., Kumar, P., & Rastogi, R. (2014). Critical review of flexible pavement performance models. *KSCE Journal of Civil Engineering, 18*(1), 142–148.
2. Saha, P., Kasibati, K., & Atadero, R .(2017). Developing pavement distress deterioration models for pavement management system using markovian probabilistic process. *Advances in Civil Engineering.*
3. Chopra, T., Parida, M., Kwatra, N., & Chopra, P. (2018). Development of pavement distress deterioration prediction model for urban road network using genetic programming. *Advances in Civil Engineering.*
4. Jeong, H., Kim, H., Kim, K. & Kim, H. (2017). Prediction of flexible pavement deterioration in relation to climate change using fuzzy logic. *Journal of infrastructure systems, 23*(4), 04017008.
5. Gupta, A., Kumar, P., & Rastogi, R. (2011). Pavement deterioration and maintenance model for low volume roads. *International Journal of Pavement Research and Technology, 4*(4), 195–202.
6. Abu-Samra, S., Zayed T., & Tabra, W. (2017). Pavement condition rating using mutliattribute utility theory. *Journal of Transportation Engineering, Part B: Pavements, 143*(3), 04017011.
7. Premkumar, L., & Vavrik, W. R. (2016). Enhancing pavement performance prediction models for Illinois Tollway system. *International Journal of Pavement Research and Technology, 9*, (1), 14–19.
8. Suryoto, Siswoyo, D. P., & Setyawan, A. (2017). The evaluation of functional performance of national roadway using three types of pavement assessments methods. *Procedia Engineering, 171*, 1435–1442.
9. Tawalare, A., & Raju, K. V. (2016). Pavement performance index for Indian rural roads. Recent trends in engineering and material science. *Perspectives in Science, 8*, 447–451.
10. Shahnazuri, H., Tutunchian, M. A., Mashayekhi, M., & Amini, A. A. (2012). Application of soft computing for the prediction of pavement condition index. *Journal of Transportation Engineering, 12*, 1495–1506.
11. Adlinge, S. S., & Gupta, A. (2013). Pavement deterioration and its causes. *International Journal of Innovative Research and Development, 2*(4), 437–450.
12. Park, S. H., & Kim, J. H. (2019). Comparative analysis of performance prediction models for flexible pavements. *Journal of Transportation Engineering, Part B: Pavements, 145*(1), 04018062.

13. Singh, A. P., Sharma, A., Misra, R., Wagle, M., & Sarkar, A. K. (2018). Pavement condition assessment using soft computing techniques. *International Journal of Pavement Research and Technology, 11,* 564–581.
14. Luo, X., Gu, F., Zhang, Y., Lytton, R., & Zollinger, D. (2017). Mechanistic empirical models for better construction of subgrade and unbound layers influence on pavement performance. *Transportation Geotechnics, 13,* 52–68.
15. Ibrahim, D. (2016). An overview of soft computing. *Procedia Computer Science, 102,* 34–38.
16. Karaşahin, M., & Terzi, S. (2014). Performance model for asphalt concrete pavement based on the fuzzy logic approach. *Transport, 29*(1), 18–27.

Land Base and Digital Elevation Model Creation Using Unmanned Aerial Vehicle

Anupoju Varaprasad, Kundangi Haritha, Shaik Syffudin Soz, and Samoju Chiranjeevi Achari

Abstract Digital elevation model (DEM) layer is most useful for the preparation of water resource applications for creating catchment area and stream order, based on the resolution of DEM the stream order and basin area are changed. Methods behind DEM satellite images, aerial photogrammetry, unmanned aerial vehicle (UAV), GPS, and so on. UAV system is a data collection platform, and as a measurement instrument, UAV gives the high-resolution DEM by the stereopsis principle is becoming attractive for many surveying applications in civil engineering. The present study has investigated the impact of DEM on basin creation and stream order of obtained resolution from UAV of study area at M.V.G.R College Engineering. In this work, a mobile app for capturing aerial images using UAVs, around an area of 63.0 acres, acquired 149 photographs at the 510 ft elevation. Resampled DEM resolution from 5 to 30 m and compared the outputs with the BHUVAN, ASTER, and USGS resolutions. Orthomosaic used as input data set to create the digital layout of study area, DEM as input for water resource applications, and followed the steps as fill, flow direction, flow accumulation, stream order, and basin by ArcGIS software. The present study created the high-resolution DEM using UAV is 0.325 m and compared with BHUVAN, ASTER, and USGS of resolutions with the help of 20 GPS as checkpoints in the campus at the time of UAV survey. The profile of UAV DEM elevation points is matching with GPS, ASTER, and USGS DEM, except 5 points, and these points are covered with trees within a distance of max 1.5 m. Also, stream order and basins are also affected with resolutions of DEM, and it was compared with 5, 10 20, and 30 m. The stream order was changed from 9 to 4, and basins are changed from 10,359 to 25. High resolution of DEM gives better details compared with medium resolution DEM for small catchment areas. Land base data was prepared with orthoimages of resolution 0.325 m. The significant layers are roads, road median,

A. Varaprasad (✉)
Assistant Professor, Department of Civil, MVGR College of Engineering, Chintalavasa, Vizianagram 535005, India
e-mail: avaraprasad@mvgrce.edu.in

K. Haritha · S. S. Soz · S. C. Achari
Students, Department of Civil, MVGR College of Engineering, Chintalavasa, Vizianagram 535005, India

buildings, playgrounds, poles, trees, water bodies, and seating benches. Finally, it was concluded that UAV applicable to the collection of qualitative data and analysis of collected data.

Keywords DEM · GIS · Unmanned aerial vehicle · Land base · Water resource applications

1 Introduction

Nowadays, there are many provisions to find the water resource applications form a digital elevation model. Creation of DEM's is very easy using unmanned aerial vehicle (UAV). In UAV-based projects, both how to collect the data and the quality of the collected data are of high importance. Methods for data capture are GPS, LIDAR, total station, aerial photogrammetry, and unmanned aerial vehicles. UAVs have become common in recent years for the best projects. UAVs play a significant role in photogrammetry and are a generic aircraft design to operate with no human pilot onboard. In this community, UAV is also named as Drone Remotely Piloted Vehicle (RPV), Remotely Operated Aircraft (ROA), Micro Aerial Vehicles (MAV), Unmanned Combat Air Vehicle (UCAV), Small UAV, Remote Control Helicopter, and Model Helicopter are often used, according to their propulsion system, altitude or endurance and the level of automation in the flight execution [1].

The latest UAV success and developments can describe by the spreading of the platform combined with the SRL Digital Cameras and Global National Satellite Systems, necessary to navigate the platform, predict the acquisition points, and possibly perform direct georeferencing. Although conventional airborne remote sensing has still some advantages and it is reducing the gap between airborne and satellite mapping applications in the resolution of images, UAV platforms are a critical alternative and solution for studying and exploring the environment, in particular, for heritage locations or rapid response applications.

The applications are rapidly growing to agriculture, scientific, commercial, recreational geographical information system, and other applications such as policing, peacekeeping and surveillance, product deliveries, aerial photographs, smuggling, and drone racing [2].

Digital elevation model is the digital representation of land surface elevation with respect to the reference datum. It is also the natural form of digital delineation of topography, used to determine terrine attributes such as elevation at any point, slope, and aspect. Terrine features like drainage basin and channel networks can be identified from the DEM's. Different software's helpful for DEM creation and also applicable, further outputs from DEM which are WEBODM, AGISOFT METADATASHAPE PROFESSIONAL, PIX4D, ARCGISPRO, and DRONE MAPPER.

Cartography is the art, science, and technology of making maps, together with their study as scientific documents and works of art as given definition by International Cartographic Association in 1973. In this cartography may be regarded as including all types of maps, plans, charts, and sections, 3D models and globes representing the Earth or any extra-terrestrial body to any scale. A map is a symbolic representation of objects, regions, or themes. It is a relatively organized presentation of a given physical space. A road map is critical for a person trying to travel in an unfamiliar place. The road map can help that person to know where this are going and plan out alternate routes to get to that destination. A better example for this is a toposheet to Civil engineers which consist of the road network, settlement, drainage network, electric lines etc. Map layout in ArcGIS allows users to create maps of spatial data quickly, and to include charts, tables, north arrows, scale bars, text, graphical primitives, and graphics files [3].

This documentation provides the procedure for digital elevation model and water resource applications, and layout representation for various features, rainwater flow path, and basin area for the study area.

1.1 Literature Review

Aleksandra [4] aimed to examine the potential of image classification and judge if it could be a satisfying alternative for point cloud classification. As per the Aleksandra, the first step in this project was capturing data with the help of phantom 3 advanced. The photogrammetric flight was prepared as per drone specifications, and objectives of the survey. After the flight, the photographs were processed with the use of Agisoft PhotoScan and Pix4D Mapper, root mean square (RMS) for control points are 6.12 cm for Pix4D and 7.91 for Agisoft, the results to this experiment.

In this project, significant steps of methodology were dense point clouds, DEMs, and orthophoto maps were generated, and point cloud classification was performed. Zietara adopted orthophoto map from Pix4D had better resolution (3.36 cm), and using this map, she processed image classification. Also, further, she performed map with ArcGIS Pro software and obtained the highest accuracy from random trees supervised classifier (76%). Manual improvements were implemented and map reclassified into two classes: ground and non-ground objects. DEM obtained from the image classification compared with the DEM obtained from the point cloud classification. For a significant part of the study area, the difference was smaller than the 0.5 m, but for both models, problematic regions appear in the proximity of buildings and trees.

Jan Komarek, Jitka Kumhalova, et al. [5] aimed to provide insight into modeling and assessment of a digital surface model of experimental plot. Jan, Jitka, and Milan had used an unmanned aerial vehicle equipped with the low-cost RGB camera. In this work acquired 251 perpendicular images per 11.5 ha of the suggested plot, taken from the altitude of 50 m above the study area. Orthomosaic resolution in set height was 0.02 m pixel1 and did registration to the coordinate system by ground control points (GCP) measured by the real-time kinematic global positioning system. For this purpose, done survey for 21 GCPs. Over 24 million georeferenced points were used in the digital surface model generation. The root mean square error (RMSE) method gives the value as 0.29 m with the number of 21 precise checkpoints measured by the GPS in real-time kinematic mode per 11.5 ha.

As Sammartano and Antonia [6], they reviewed that for the use of terrestrial and aerial sensor better technology is Geomatics and it is giving significant support and new potentialities in terms of multi-scale precision, quickness and cost-cutting. Mapping and 3D data products, all these can be derived from aerial acquisitions in large scale. The paper is mostly purposed to examine the use of tools aimed to generate the DEM from DSM obtained from UAV flights. In kinds of literature, many applications concern the point cloud data generation from aerial photogrammetry. Several different filtering approaches and algorithms (filtering points along with density, direction, and slope) are used to derive bare-earth, but in the test case, the high level of detail of objects, together with the complexity of high slope of ground impose some adaptation. The test is included in a decision-making process concerning the promotion of alpine landscape led through a project of the sustainable path of the railway track, achieved by a simple multi-criteria analysis performed by geographic information system tools.

As per Samira, Even, Raheeb [7], days where swarms of unmanned aerial vehicles will occupy our sky easily and it is fastly approaches to customer due to cost-efficient and reliable with visibility looking in small aerial vehicles. Hence, the demand increases for the use of such vehicles in an abundance of civil applications. Governments and industry alike have been heavily investing in the development of UAVs. As such, it is essential to understand the characteristics of the network with UAVs to enable the incorporation of multiple, coordinate aerial vehicles into the air traffic reliably and safely. From this survey reports that communications and network viewpoint were used over the period 2000–2015 for civil applications. For this purpose, surveyed and qualify the quality of service requirements, network relevant mission parameters, data requirements, and the minimum data to be transmitted over the network. And also elaborate general networking-related requirements such as connectivity, adaptability, safety, privacy, security, and scalability. It reports experimental results from many projects and investigate the suitability of existing communication technologies for supporting reliable aerial networking.

1.2 Summary from Literature Review

As per the above literatures, different people proposed differently about how UAV used to their projects and solved applications, it deals more about the creation of DSM and DEM. Some of them are examined the capability of drone in aerial photogrammetry and software tools used the creation of DEM and also examine the resolution potentiality of DEMs with standard DEMs. Some of them are applied aerial photogrammetry by unmanned aerial photogrammetry to solve civil-related applications and the creation of water resource applications. Observed that UAVs are capable of doing a low-cost project with more efficient. There are software's like Agisoft,

Pix4D, ArcGIS Pro applied to create digital elevation model. It concludes that UAVs have different applications to solve in different study areas.

2 Materials and Methods

2.1 Drone Deploy

Drone deploy is an integrated solution used to create aerial images and 3D models. Using their mobile app—which is available in both Android and IOS—it is possible to transform a range of DJI 4pro. Drones deploy is a reliable and powerful mapping tool. Drone deploy facilitates automated flights from take-off and right through to landing while automatically capturing images that can be uploaded to their software for processing. This image can be used to create a range of visualized data from 2D maps to 3D models.

2.2 Agisoft Metashape Professional

Agisoft Metashape is a stand-alone software product that performs photogrammetric processing of digital images and generates 3D spatial data to be used in GIS applications, cultural heritage documentation, and visual effects production as well as for indirect measurements of objects of various scales.

Wisely implemented digital photogrammetry technique enforced with computer vision methods results in smart automated processing system that, on the one hand, can be managed by a new-comer in the field of photogrammetry, yet, on the other hand, has a lot to offer to a specialist who can adjust the workflow to numerous specific tasks and different types of data. Throughout various case studies, Agisoft Metashape proves to produce quality and accurate results.

2.3 ArcGIS

ArcGIS is a cloud-based mapping and analysis solution tool. Use it to make maps, analyze data, and to share and collaborate. ArcGIS is a geographic information system for working with maps and geographic information. It is used for creating and using maps, compiling geographic data, analyzing mapped information, sharing and discovering geographic information, using maps and geographic information in a range of applications, and managing geographic information in a database. An ESRI company develops ArcGIS.

2.4 Study Area

The present study area chosen is the property of M.V.G.R College of Engineering. This college is located at Vizianagaram, the latitude of Vizianagaram district is N 18° 6′ 23.9724″ and longitude is E 83° 23′ 43.9944″. The total covered area of building 63.0 acres (0.25 km^2) and total perimeter 3 km and shown in Fig. 1.

3 Methodology

The methodology is explained in the form of a flowchart as shown in Fig. 2. The UAV used for this analysis is 'Phantom 4pro' produced by unmanned aerial vehicle system. The aircraft is an auto piled system enables the aircraft to follow the predefined path, equipped with GPS receiver so that one may cover the whole test site with a suitable amount of imagery. Using this UAV has operated with drone deploy, a mobile app collected the 149 images of 0.23 km^2 for study area at the flight height of 510 ft with side overlap 65% and front overlap 75%. This imagery of 149 photographs have applied to Agisoft Metashape Professional is a stand-alone software product that performs photogrammetric processing of digital images and generates 3D spatial data to be used in GIS applications, cultural heritage documentation, and visual effects production as well as for indirect measurements of objects of various scales. In this study, it includes various steps to produce DEM as output and initially adding images, align photographs, dense point cloud, mesh creation, build texture, the creation of

Fig. 1 Study area

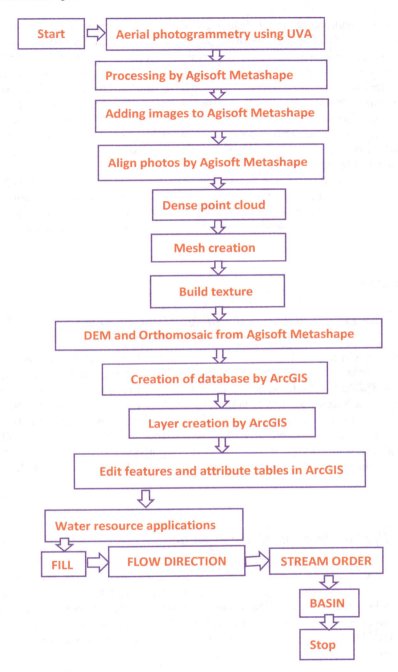

Fig. 2 Methodology flow chart

orthoimage, and finally digital elevation model. After adding images, it will check the tie point of photographs to overlap one another to create a complete image, i.e., align photographs. There are two options of dense point cloud classification: ground points and the rest, manual selection of a group of points to be placed in a particular class from the standard list known for LIDAR data. Dense cloud point's classification opens the way to customize build mesh step. After mesh creation and texture building, it gives orthomosaic as one of the outputs which are helpful in the creation of a database for the study in cartographical representation using ArcGIS catalog, ArcGIS map, and ArcGIS toolbox. The orthomosaic image of the study area is shown in Fig. 3(a).

DEM is another output for the Agisoft Metashape and is useful for the creation of water resource applications (fill, flow direction, flow accumulation, stream order, and basin) and these are also obtained using the ArcGIS toolbox tool hydrology from spatial analyst tool. The DEM is created using this and shown in Fig. 3(c).

4 Results

4.1 The Layout of M.V.G.R College of Engineering

The database is a set of features shown in Fig. 3(b), showing a boundary, buildings, road networks, trees, poles, benches, water bodies, and playgrounds. All these are classified in attribute table as per their purpose and existing in nature and also assigned symbols and colors are nearly to understand by the viewer by symbology. These are digital output representations from ArcGIS. Table 1 gives and explains details about the features type and numerical data, features some existing in M.V.G.R College. Also gives the feature type polyline or point or polygon are useful for geometrical measurement (length, perimeter, area, and elevation) of the database (Fig. 4; Table 1).

4.2 Digital Elevation Model as Output

Using unmanned aerial vehicle for photogrammetry total of 149 images is obtained at the 510ft flight. This imagery helps in obtaining DEM by Agisoft Metashape, process involved in the creation of DEM explained in the methodology chapter in detail. Obtained DEM is of resolution (0.325, 0.325 m) and compared with the USGS, BHUVAN DEMs to check the accuracy of obtained DEM and also done resampling of DEM to 5–30 m.

Digital elevation model having applications in water resource engineering in this study from DEM obtained secondary output related to the hydrology of study area. By the ArcGIS software, DEM as the input obtained fill, flow direction, flow accumulation, stream order, and basin. Based on these tools created layer and combined layers shown with DEM, Stream order and basin of the study area in Fig. 4.

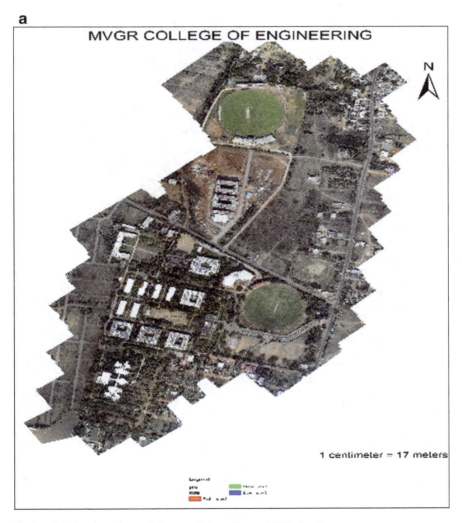

Fig. 3 **a** Mosiac drone image **b** Layout of the college **c** DEM of college

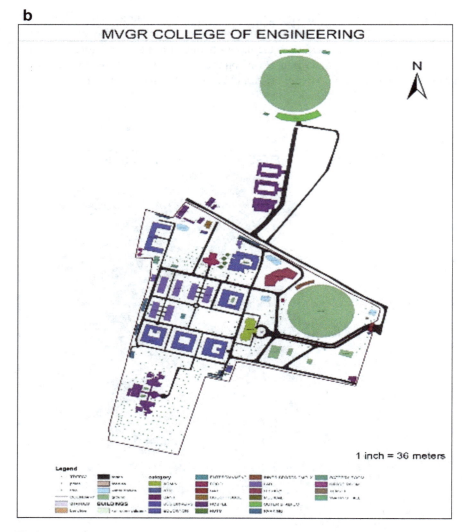

Fig. 3 (continued)

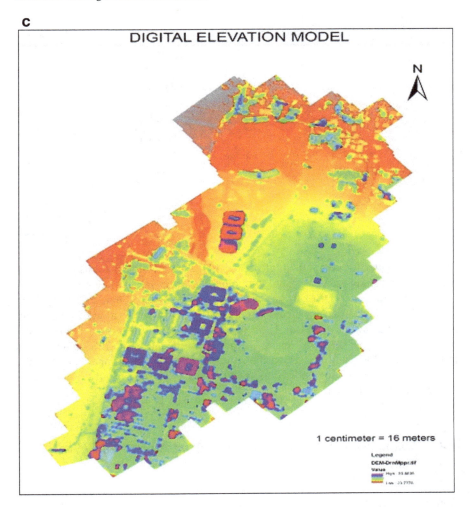

Fig. 3 (continued)

4.3 Comparison of DEMs

From pixel values, the table above chart compares the pixel value of different DEMs from chat obtained that all DEMs similarity in pixel values at a major rate at some points of DEMs has deviations with compared with other DEMs this is the major reason for the resolution changes from DEM to DEM. Comparison done by regression gives the percentage of matching. Depends on the flight height picture resolution may also change effects output resolution of the digital elevation model. X-axes are representing the name of the checkpoint and Y-axes representing pixel values. From the table of comparison, BHUVAN DEM has negative pixel values, so this DEM

Table 1 Showing numerical data of existing features of M.V.G.R

S.No.	Name of feature	Type of feature	No. of features
1	Buildings	Polygon	59
2	Playgrounds	Polygon	12
3	Waterbodies	Polygon	4
4	Median	Polygon	1
5	Benches	Polygon	14
6	Statues	Polygon	5
7	Drainage	Polygon	4
8	Roads	Polygon	2
9	Boundary	Polygon	1
10	Trees	Point	253
11	Poles	Point	7

does not consider for comparison of obtained DEM. Elevation patterns of different location in the study area compared with other elevation values shown in Fig. 5.

At PR3 of obtained DEM from UAV survey have observed that deviation in the pixel value this because of a greater number of trees at that location of GPS point. At C12 of resampling, DEM of 10 m have observed that deviation due to signal tower and the errors observed due to the not proper location of GPS points shown in Fig. 6. PR4, PR7, and PR10 also show the deviations in the comparison chart the reason for these deviations also same as PR3 and C12. Rest of all GPS points is similar for all the DEMs.

4.4 Comparison Between UAV DEM from Agisoft and USGS DEM

Comparison chart prepared for UAV DEM, USGS DEM, and USGS N18 are shown in Fig. 7. The comparison clears that pixel values of both the DEMs are similar with 80–90% regression. Also, clears that obtained from UAV DEM has nearly linear shape has minimum differences in pixel value this because of the study area is likely plain terrain in nature. USGS may use higher flight height compared to used flight height to cover the maximum area of the world. Comparison acquires the significant factor as 0.63 m, which is less than 1 m; hence, DEM is useful for further operations.

Same UAV DEM and n18 DEM are similar in pixel values and also compared between them with regression as 85 and 90%. Moreover, it has a significant factor 0.24 m which is less than 1 m based on this comparison UAV DEM has a significant feature in various applications (Fig. 7).

Table 2 Pixel values of different DEMs

Name	Elevation by GPS	Res5m	Res10m	Res20m	Res30m	N18	Original	BHUVAN	USGS	ASTER
c02	66.629	30.636	30.545	30.453	30.453	61	30.636	−10	59	59
c03	66.22	31.002	31.002	31.093	30.819	60	31.002	−7	60	60
c11	68.387	35.209	35.025	37.678	39.141	67	35.117	−1	61	61
c12	66.282	33.197	33.288	34.02	33.471	61	33.197	−5	64	64
c14	68.878	35.574	35.574	36.214	35.574	62	35.574	−4	62	62
c15	68.959	36.489	36.58	36.397	38.226	66	36.489	−3	61	61
c16	68.579	35.849	35.849	36.489	36.123	66	35.849	−6	60	60
o6	66.11	33.654	33.563	33.654	33.38	63	33.654	−9	58	58
o7	65.067	32.374	32.099	31.916	32.008	64	32.191	−11	61	61
pr3	21.377	34.843	32.557	33.105	33.288	63	33.563	−10	63	63
pr4	24.06	32.099	33.288	35.117	31.642	65	31.459	−9	62	62
pr7	31.708	31.734	31.825	31.459	31.734	59	31.551	−9	61	61
pr10	46.29	34.477	34.568	34.477	39.324	63	34.477	−7	59	59
pr11	48	35.026	35.026	38.318	35.483	59	35.117	−7	55	55
pr14	54.812	35.574	35.574	35.666	35.391	60	35.483	−6	61	61
pr15	55.937	35.849	35.757	35.757	36.123	64	35.849	−7	61	61
pr23	62.364	33.288	33.563	36.763	33.014	61	33.654	0	68	68
pr24	62.143	32.648	32.831	35.849	33.563	66	32.648	−7	71	71
pr25	62.205	32.191	32.282	32.557	33.38	64	32.191	−6	60	60
pr27	64.205	33.654	33.745	33.654	33.837	65	33.745	−5	67	67

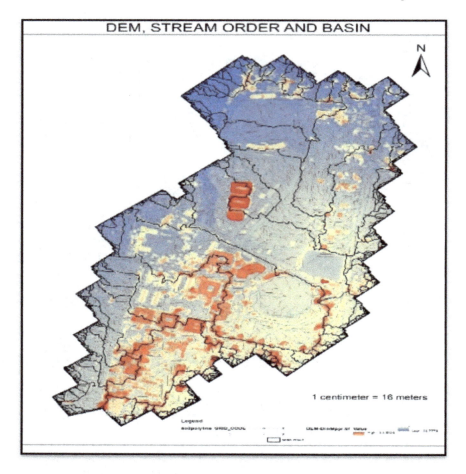

Fig. 4 DEM, stream order, and basin of study area

4.5 Water Resource Applications as a Result

Digital Elevation Model is generated from photographs using the unmanned aerial vehicle with 0.325 m (x = 0.326 m, y = 0.326 m) as resolution. Using this UAV DEM created the water resource application by using ArcGIS toolbar where spatial reference and hydrology tools play a significant role in hydrology and it has different applications of creating a digital database for the fill, flow direction, flow accumulation, stream order, and basin for the study area of 0.23 km² (Fig. 8).

Stream Orders from Resampled UAV DEMs, USGS, BHUVAN, and ASTER
However, the present study shows stream order and basin as the primary output for water resource applications. Stream order gives the maximum degree of the flow of rainwater to lower elevation. Basin gives the difference in elevation from lower to

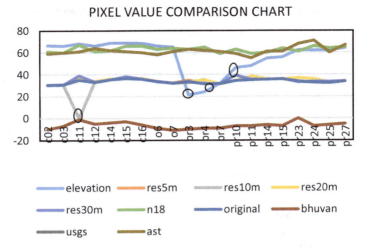

Fig. 5 Elevation patterns of different locations comparing with other elevation values

Table 3 Stream order and basin counts

DEM file	Stream order count	Basin count
Res5m	6	1–251
Res10m	5	1–98
Res20m	4	1–45
Res30m	4	5–25
USGS	4	1–30
AST	5	6–27
N18	5	2–37
UAV DEM	9	1–10,359

Fig. 6 Error due to the not proper location of GPS points

high and shows no. of basins side by side with elevation differences whereas from ArcGIS gives the difference pixel value with the help of grid code. Created hydrology applications already mentioned and are shown in the above methodology chapter. Also, combined basin and stream order layer with UAV DEM of the study area.

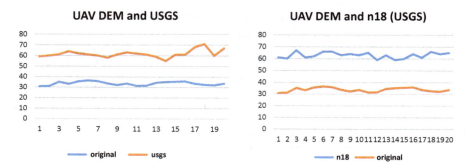

Fig. 7 Comparison chart UAV DEM-USGS DEM and UAV DEM-N18

Basins from Resampled UAV DEM, USGS, BHUVAN, and ASTER Stream order is a method of assigning a numeric order to links in a stream network. Basin explains about the drainage network. Table 3 gives stream order count and basin count, for some basins count reduced to obtain better visibility. The basin of different DEMs are shown and at different resolution shown in Fig. 8.

Stream Order and Basin Combined with UAV DEM

Shown in Fig. 9 is obtained from the conversion of basin (polygon) to polyline and merged all the polylines into a single polyline. By adding stream order, it will be able to identify the flow of drainage water and capable of locating the places where the water gets stacked in the study area.

5 Summary and Conclusions

Land base and digital elevation model of the study area at M.V.G.R College of Engineering have been developed from an aerial survey using unmanned aerial vehicle. There are many ways to collect the data for DEM preparation, but UAV has proven that it has a higher potential in creating accurate digital elevation model like applications water resource. Also, DEM has proven that it would give quality in water resource application products. Even the resolution of DEM also affects the quality of stream order and basin. Along with DEM, orthomosaic has given the digital output for the study area. Orthomosaic has an application of high-resolution creation database compared with satellite images, whereas, in this study, we aimed to prepare water recourse applications from the DEM along with the mapping of land features of the study area from Orthomosaic. Obtained DEM resolution of (0.325, 0.325 m) it is more than enough for quality Fig. 4 products of hydrology. However, there are some negligible deviations in mapping this because of hand on work to draw the shapes of existing features in ArcGIS.

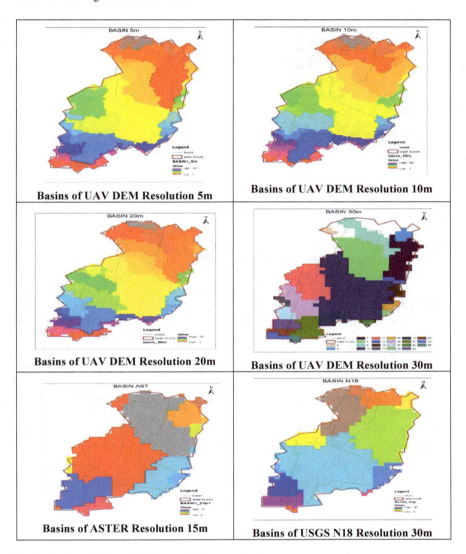

Fig. 8 Basins of different DEMs and different resolutions

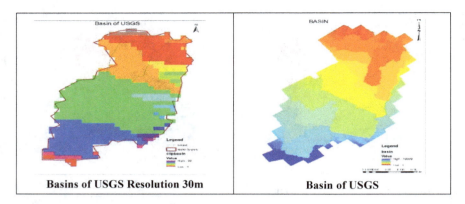

Fig. 8 (continued)

From the present study, it has resulted in a DEM with a high resolution of 0.325 m using Unmanned Aerial Vehicle. It is compared with BHUVAN, ASTER, and USGS of resolutions of 30 m with taking 20 GPS points in the campus at the time of UAV survey. The profile of UAV DEM elevation points is matching with GPS, ASTER, and USGS DEM, except 4 points, these points are covered with trees within a distance of max 1.5 m. Also, stream order and basins are effective with resolutions of DEM, and it is compared with 5, 10, 20, and 30 m. The stream order is changed from 9 to 4, and basins are changed from 10,359 to 25. High resolution of DEM gives better details compared with medium resolution DEM for small catchment areas. Land base data was prepared with orthoimages which were created with UAV images of resolution 0.325 m. The significant layers are roads, road median, buildings, playgrounds, poles, trees, water bodies, and seating benches. It was concluded that unmanned aerial vehicle is also applicable for the collection of qualitative data and analysis of collected data.

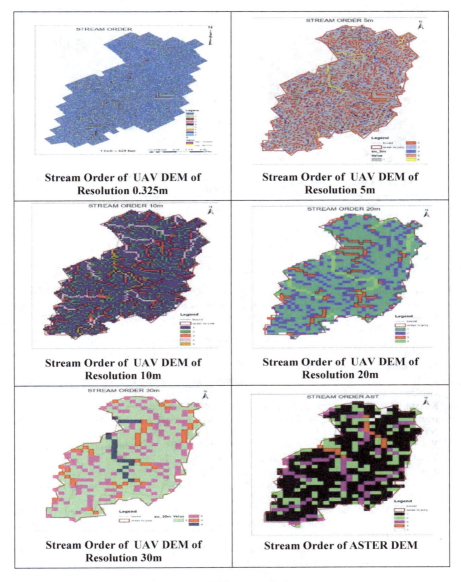

Fig. 9 Stream order of different DEMs and different resolutions

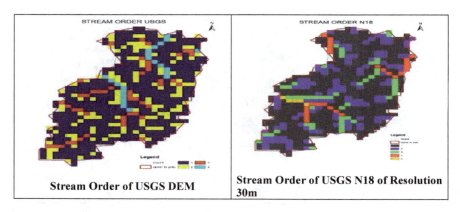

Fig. 9 (continued)

References

1. Ahmad, A., Tahar, K. N., Udin, W. S., Hashim, K. A., Darwin, N., Hafis, M. et. al. (2013). Digital aerial imagery of unmanned aerial vehicle for various applications. In *2013 IEEE International Conference on Control System, Computing and Engineering* (pp. 535–540). EEE. https://doi.org/10.1109/iccsce.2013.6720023.
2. Everaerts, J. (2008). The use of unmanned aerial vehicles (UAVs) for remote sensing and mapping. *The International Archives of the Photogrammetry, Remote Sensing and Spatial Information Sciences, 37*(2008):1187–1192.
3. Lanfear, K. J. (1992). Water resource applications of geographic information systems by the US Geological Survey. *Journal of Contemporary Water Research and Education, 87*(1), 2.
4. Zietara, A.M. (2017). Creating digital elevation model (DEM) based on ground points extracted from classified aerial images obtained from Unmanned Aerial Vehicle (UAV) (Master's thesis, NTNU).
5. Komarek, J., Kumhalova, J., & Kroulik, M. (2016). Surface modelling based on unmanned aerial vehicle photogrammetry and its accuracy assessment. In *Proceedings of the International Scientific Conference Engineering for Rural Development*, Jeglava, Latvia (pp. 25–27).
6. Sammartano, G., & Spanò, A. (2016). *DEM generation based on UAV photogrammetry data in critical areas.* In GISTAM (pp. 92–98).
7. Hayat, S., Yanmaz, E., & Muzaffar, R. (2016). Survey on unmanned aerial vehicle networks for civil applications: A communications viewpoint. *IEEE Communications Surveys & Tutorials, 18*(4), 2624–2661.
8. Meynen, E. (1973). *Multilingual dictionary of technical terms in cartography: Dictionnaire Multilingue de Termes Techniques Cartographiques.*

Multiphase Modelling of Orifice Cavitation for Optimum Entrance Roundness

V. R. Greeshma and R. Miji Cherian

Abstract Cavitation is a common problem in pumps, control valves and orifices; causing serious wear, tear and damages. Under the wrong condition, cavitation reduces the components lifetime drastically. Cavitation occurs when local static pressure in a fluid reaches a level below the vapour pressure of the liquid at actual temperature. This work aims to study the cavitating flow through a circular orifice using computational fluid dynamics and to find out optimum shape of orifice in terms of entrance roundness to minimize cavitation. Fluid model of sharp-edged circular orifice is created in CFD, and the obtained result is compared with that of available experimental result. After calibration and validation, the fluid models created for circular orifice is modified for different entrance roundness values. The influence of entrance roundness on cavitation characteristics is analysed and the optimum value of entrance roundness for minimizing cavitation is obtained.

Keywords Cavitation · Cavitation number · Computational fluid dynamics · Optimum entrance roundness

1 Introduction

Cavitation phenomena are ubiquitous in flows which contain large velocity differences between a liquid and a solid boundary, Chen et al. [1]. Cavitation is the rapid evolution of vapour or gas bubbles within fluid due to very low pressure in fluid flow, or due to a rise in temperature, which raises the saturation pressure. The sudden formation of bubbles causes rapid variation in pressure within incompressible fluids, which in turn can cause severe mechanical damages. The small bubbles of water

---

V. R. Greeshma (✉)
Government Engineering College Trichur, Thrissur, Kerala, India
e-mail: greeshmavram94@gmail.com

R. Miji Cherian
Department of Civil Engineering, Government Engineering College Trichur, Thrissur, Kerala, India
e-mail: mijicpaul@gectcr.ac.in

vapour will move with the water from low pressure area into areas of higher pressure. The damage caused by cavitation happens when these water vapour bubbles implode due to that higher pressure. The collapsing bubbles send out little shock waves. These waves can cause small bits at the surface of the pump or pipe to break away. The erosion of the material can actually reach a point where a hole can develop in the impeller or piping. Therefore, it is critical to understand where cavitation is likely to occur and how bad it is likely to become. Since it is difficult to initiate and visualize cavitation experimentally and the cavitation is potentially damaging, it is critical to simulate the cavitation process.

Cavitation can occur after a sudden contraction in a conduit and can cause volumetric oscillations of bubble nuclei in the flow due to ambient pressure change [2]. Nurick [3] conducted experimental study on orifice cavitation characteristics over a range in back pressure and L/d for both circular and rectangular orifices, where L is the length of the orifice and d is the diameter of the orifice. A variety of single orifices, including those fabricated from lucite, stainless steel and aluminium, were used to determine cavitation characteristics. Critical flow conditions can be calculated using a relatively simple correlation between cavitation number and discharge coefficient. Cavitation occurs in sharp-edged orifices leading to either hydraulic flip or reattachment, depending on the flow conditions and orifice L/d. Yang et al. [4] conducted simulation-oriented researches on orifice plate cavitation phenomenon by using ANSYS FLUENT due to its high resolution comparing to experiments. Standard orifice plates based on ASME PTC 19.5-2004 were chosen and modelled in the study with the diameter ratio from 0.2 to 0.75. To investigate the dependency of discharge coefficient on the cavitation number, steady-state studies were conducted for each diameter ratio (ratio of orifice diameter to pipe diameter) at the cavitation number varied roughly from 0.2 to 2.5. They concluded that 3D simulation can be replaced by 2D axisymmetric simulation in favour of time and cost efficiency. They created a regression polynomial that can be used to estimate the discharge coefficient at different diameter ratios. Tukiman et al. [5] used commercial CFD software to predict the flow features in the orifice flow metre. The *k-epsilon* model was chosen for the turbulence model and the numerical accuracy was set to first order. All the discretized equations were solved in a segregated manner with the Semi-Implicit Pressure-Linked Equation (SIMPLE) algorithm. The results agreed well with the published data in terms of flow pattern, velocity profiles and pressure profile. They concluded that the CFD technique can be used as an alternative and cost-effective tool towards the replacement of experiments.

Cavitation become a major threat to many of the civil engineering structures or hydraulic components, especially closed conduits such as pipelines, valves, pumps and storage tank. Hence, it become essential to find a proper solution to overcome the problems arises due to formation of large cavities within the conduits. Orifice is a conduit having restricted area when compared to the upstream or and downstream section of the pipeline. When fluid passes through the orifice, its pressure builds up slightly upstream of the orifice. But as the fluid is forced to converge, the flow velocity increased and fluid pressure decreased. Thus, the orifices are highly potential to cavitation and cavitation damages. The effect of abrupt change in cross section and gradual change in cross section definitely vary.

2D axisymmetric simulation is superior over 3D simulation in terms of time and cost efficiency. The geometry of orifices has great influence in the cavitation characteristics. Researchers have conducted studies on influence of aspect ratio, diameter ratio and radius of orifice on cavitation characteristics. Investigations on the influence of entrance roundness of orifice on cavitation characteristics are not conducted so far. Hence, the present study focuses on the 2D analysis of cavitating flow in orifices and on optimization of entrance roundness of orifice to minimize cavitation using ANSYS FLUENT.

2 Methodology

ANSYS FLUENT is used for the development of 2D fluid model and the simulation of cavitation. Fluid model of circular orifice of sharp edge is created and the obtained result is compared with the experimental result available in the literature. After comparing the results, the fluid model created for circular orifice is modified for different entrance roundness values. For each fluid model created, the simulations are carried out for different boundary conditions. By comparing the simulation results obtained, an optimum entrance roundness value, which has minimum cavitation, is identified.

2.1 Experimental Data from Literature

The experimental study conducted by Nurick [3] is considered for this study. The geometric parameters of circular orifice are $D/d = 2.88$ and $L/d = 5$, where D, d and L are inlet diameter, orifice diameter and orifice length, respectively. Figure 1 shows the geometric parameters for the circular orifice. The flow is axisymmetric.

Figure 2 shows all the boundary conditions utilized in this work. Uniform inlet and outlet static pressure are adopted as boundary conditions. The exit pressure is fixed as 95 kPa and the upstream pressure varied from 250 to 1500 kPa. The parameters to validate the model are the cavitation number (σ) and the discharge coefficient (C_d):

Cavitation number $\sigma = \frac{P_i - P_v}{P_i - P_o}$

Discharge coefficient $C_d = \frac{\dot{m}_{actual}}{\dot{m}_{ideal}} = \frac{\overline{V_b}}{\sqrt{\frac{2(P_i - P_o)}{\rho_l}}}$ $C_d = \frac{\dot{m}_{actual}}{\dot{m}_{ideal}} = \frac{\overline{V_b}}{\sqrt{\frac{2(P_i - P_o)}{\rho_l}}}$ and $C_d = C_c \sqrt{\sigma}$

where P_i, P_o and P_v are the upstream (inlet), vapour and exit (outlet) pressure, respectively; C_c the contraction coefficient of orifice, the average velocity at the exit, the liquid density and the mass flow.

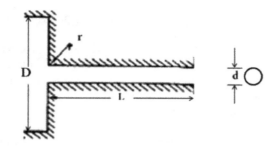

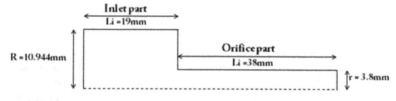

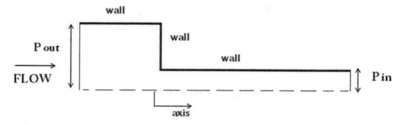

Fig. 1 Parameters and present geometry for the circular orifice [6]

Fig. 2 Boundary conditions [6]

The geometry of orifices is modified for r/d ratios of 0.04, 0.08, 0.1, 0.12 and 0.16, where r is the radius of curvature provided at orifice entrance and d is the diameter of the orifice. Then, similar to the sharp-edged orifice, simulations are carried out for each orifice with different r/d ratios for all the boundary conditions.

2.2 Software Used

ANSYS Workbench 16.0 is an industrial CFD software programme used for the study. The fluid flow modelling and analysis accomplished in ANSYS FLUENT solving the Navier–Stokes equations. Numerical simulation based on the finite volume method was performed to solve the continuity and momentum equation.

The continuity equation for a compressible fluid may be written as follows.

$$\frac{\partial \rho}{\partial t} + \frac{\partial (\rho U_t)}{\partial x_t} = 0 \tag{1}$$

Here, ρ is the fluid density and U_t is the fluid velocity x_t direction.
The momentum equation is

$$\frac{\partial (\rho U_t)}{\partial t} + \frac{\partial (\rho U_t U_j)}{\partial x_j} = \frac{\partial p}{\partial x_t} + \frac{\partial}{\partial x_j}\left[\mu\left[\frac{\partial U_t}{x_j} + \frac{\partial U_j}{\partial x_j} - \frac{2}{3}\frac{\partial U_k}{\partial x_k}\delta_{tj}\right]\right] + \rho g_t + F_t \tag{2}$$

The first term at the left hand side describes the variation of the fluid momentum in time; the second term describes the transport of the momentum in the flow. The first term on the right hand side describes the effect of gradients in the pressure p; the second term, the transport of momentum due to the molecular viscosity; the third term, the effect of gravity g; and in the last term, F_t lumps together all other forces acting on the fluid.

3 Results and Discussion

Pressure-driven cavitating flow through circular orifices of various entrance geometries are simulated. In all simulations, the working fluid is water whose liquid and vapour densities are taken as 1000 and 0.02558 kg/m³, respectively. The viscosity of water and water vapour are 0.001 kg/m s and 1.26×10^{-6} kg/m s, respectively. The pressures at inlet and outlet boundaries are specified. The outlet pressure is fixed as 95 kPa and inlet pressure varied between 250 and 1500 kPa. The simulations are repeated for all the boundary conditions with circular orifice of various entrance roundness conditions of $r/d = 0.0, 0.04, 0.08, 0.10, 0.12$ and 0.16, where r is the radius of curvature provided at orifice entrance and d is the diameter of the orifice. The pipe is modelled as solid wall with no slip condition. The k-epsilon model was chosen for the turbulence model and the numerical accuracy was set to first order. All the discretized equations were solved in a segregated manner with the Semi-Implicit Pressure-Linked Equations-Consistent (SIMPLEC) algorithm. The simulation was run until the residual of the pressure and velocities were less than 0.001.

Grid independence study is conducted for a specific case with an inlet pressure of 250 kPa. For 82,829 mesh elements the mass flow rate converged, thus the same number of elements is chosen for simulations of the orifice with various boundary conditions. Similarly, for all the orifices with different entrance geometries, grid independence studies are carried out and number of grid elements for simulations are selected (Table 1).

Table 1 Inlet pressures applied and corresponding cavitation numbers

Inlet pressure (kPa)	Cavitation number
250	1.59
265	1.54
285	1.48
300	1.45
320	1.41
350	1.36
400	1.30
500	1.23
600	1.18
750	1.14
1000	1.10
1500	1.07

3.1 Calibration and Validation

For the circular sharp-edged orifice, simulations are carried out for all the boundary conditions and a graph between discharge coefficient and cavitation number is plotted. Figure 3 compares the current predicted discharge coefficient values with Nurick correlation. It can be clearly seen that the model is in agreement with the Nurick correlation.

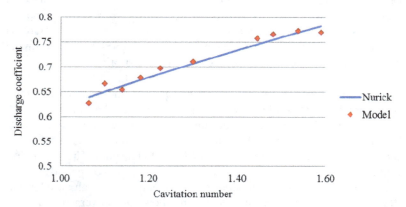

Fig. 3 Orifice cavitation: comparison of cavitation model predictions with Nurick correlation

3.2 Effect of R/d Ratio

Circular orifices with six different geometries are simulated. The entrance roundness of the orifice analysed are $r/d = 0.0, 0.04, 0.08, 0.1, 0.12$ and 0.16. From the simulation results, minimum pressure along the orifice axis and vapour volume fraction for every orifice, simulated with a set of twelve different boundary conditions, are analysed for optimizing the roundness in order to obtain minimum cavitation.

3.2.1 Minimum Pressure Along the Orifice Axis

For all the simulations pressure variations along the orifice axis are plotted. Figure 4 displays the pressure variation (Y-axis) along the axis of sharp-edged orifice (X-axis) for all cavitation numbers evaluated. For cavitation number 1.59 vena contracta can be clearly identified with a sudden dip in the curve. Further decrease in the cavitation number causes increase in the length of low pressure region along the axis and the pressures is close to vapour pressure of water at 300 K. From Fig. 4, it can be inferred that as the cavitation number decreases, the length of low pressure region along the axis increases.

In order to check whether this trend exists in other geometries adopted for the study, pressure variations for all r/d ratios are analysed. When the orifice entrance roundness is $r/d = 0.04$, for cavitation numbers $\sigma = 1.59$ and $\sigma = 1.54$ there is sharp dip in the curves indicating the vena contracta. But these dips do not reach a low pressure close to vapour pressure of water. For cavitation number $\sigma = 1.48$ and lower, the minimum pressure is close to vapour pressure of water and the length of low pressure region increases along the axis as cavitation number decreases.

Orifice with $r/d = 0.10$ the same trend as that of $r/d = 0.0$ and $r/d = 0.04$ is observed. For first four cavitation numbers, the curves are similar with a dip indicating vena contracta formation. But in these four cases, there is no chance for cavitation inception along the orifice axis since the minimum pressures along the axis are very higher than vapour pressure of water at 300 K. From $\sigma = 1.41$ onwards, the minimum pressure is close to vapour pressure of water.

The minimum pressure observed along the orifice axis for all the geometrical shapes are provided in Table 2. As cavitation number decreases the minimum pressure observed along the orifice axis decreases and after reaching a pressure near the vapour pressure there is no further decrease observed for all the orifices irrespective of the entrance geometries and in no case the minimum pressure is less than the vapour pressure of water, i.e.; there is no chance for cavitation is identified along the axis even in the sharp-edged orifice for the cavitation numbers considered. This proves that there is no column separation in any of the cases.

For cavitation numbers $\sigma = 1.59, 1.54, 1.48$ and 1.45 as r/d ratio increases minimum pressure along the axis is increasing. This indicate that the chance of cavitation along the axis (column separation) goes on reducing as the r/d ratio increases, for all the geometries and inlet pressures considered this trend exists. The minimum

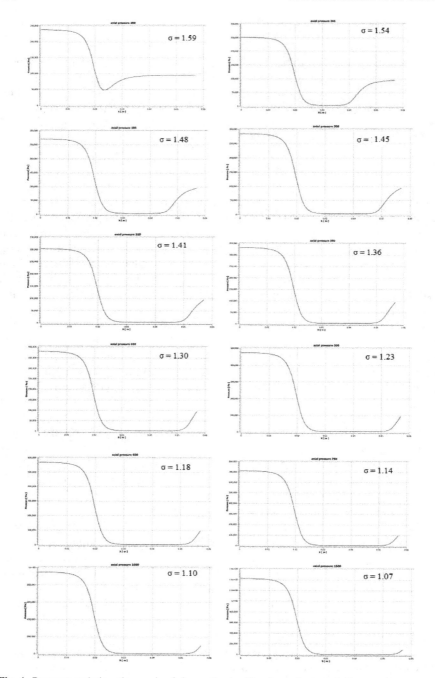

Fig. 4 Pressure variation along axis of sharp-edged orifice for various cavitation numbers

Table 2 Variation of minimum pressure with cavitation number and orifice entrance roundness

Cavitation number	Minimum pressure along the axis, P_{min} (kPa)					
	$r/d = 0.0$ (sharp-edged orifice)	$r/d = 0.04$	$r/d = 0.08$	$r/d = 0.10$	$r/d = 0.12$	$r/d = 0.16$
1.59	48.020	68.234	90.593	92.420	95	95
1.54	3.671	62.256	89.247	91.900	95	95
1.48	3.580	5.701	84.832	89.735	95	95
1.45	3.575	3.994	74.231	86.779	95	95
1.41	3.574	3.691	3.632	25.888	95	95
1.36	3.555	3.642	3.594	3.949	3.613	3.601
1.30	3.552	3.650	3.600	3.568	3.604	3.594
1.23	3.550	3.604	3.612	3.560	3.612	3.590
1.18	3.550	3.582	3.625	3.556	3.620	3.593
1.14	3.552	3.580	3.632	3.557	3.629	3.593
1.10	3.556	3.578	3.640	3.559	3.644	3.600
1.07	3.548	3.588	3.678	3.562	3.676	3.605

pressure is very near to vapour pressure in all the cases when the cavitation number is less than 1.36 (inlet pressure greater than 350 kPa). The smallest value of minimum pressure corresponding to a particular cavitation number is observed for sharp-edged orifice.

For cavitation numbers $\sigma = 1.59$, 1.54, 1.48 and 1.45 as r/d ratio increases minimum pressure along the axis is increasing. This indicate that the chance of cavitation along the axis (column separation) goes on reducing as the r/d ratio increases, for all the geometries and inlet pressures considered this trend exists. The minimum pressure is very near to vapour pressure in all the cases when the cavitation number is less than 1.36 (inlet pressure greater than 350 kPa). The smallest value of minimum pressure corresponding to a particular cavitation number is observed for sharp-edged orifice.

In all the orifices with different geometries, as the cavitation number decreases, a sudden drop in the minimum pressure is observed and the cavitation number corresponding to the sudden drop is not constant but decreases with increasing r/d ratio. For $r/d = 0.10$ onwards, it is clear that the sudden drop in the minimum pressure is observed at cavitation number 1.36. For all the orifices irrespective of the entrance roundness, the minimum pressure will be close to vapour pressure when the cavitation number is a lower value.

The graphical representation of the table is given in Fig. 5. The curves showing the variation of minimum pressure along the axis with decreasing cavitation number for circular orifices with r/d ratio 0.12 and 0.16 are overlapping.

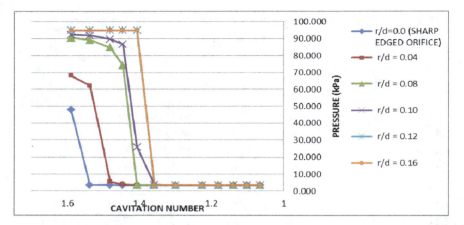

Fig. 5 Variation of minimum pressure along the axis with cavitation number and orifice entrance roundness

4 Conclusions

Pressure-driven cavitating flow through circular orifices of various entrance geometries are simulated for a set of 12 boundary conditions. Through the validation study, by comparing the curve between cavitation number and discharge coefficient of simulation and experiment results, it is appropriate and relatively accurate to implement the *k-epsilon* and Schnerr-Sauer model to simulate the cavitation flow.

The *y* analysing the results of the simulation, it can be concluded that with increasing *r/d* ratios the system become less susceptible for cavitation. With increasing inlet pressure, the chance for occurrence of cavitation increases in all the circular orifices irrespective of the entrance roundness. The conclusions from the simulation results are listed below:

- As cavitation number decreases the minimum pressure observed along the orifice axis decreases and after reaching a pressure near the vapour pressure there is no further decrease observed for all the orifices irrespective of the entrance geometries.
- In no case the minimum pressure along the orifice axis is less than the vapour pressure of water at 300 K. This proves that there is no column separation in any of the cases.
- For *r/d* = 0.10 onwards, it is clear that the sudden drop in the minimum pressure is observed at cavitation number 1.36.
- For all the orifices irrespective of the entrance roundness, the minimum pressure will be close to vapour pressure when the cavitation number is less.
- The value of minimum pressure is observed to be lowest for sharp-edged orifice.
- With increasing roundness of the orifice entrance, it is possible to reduce the chance for cavitation inception.

- For lower cavitation numbers, there is more chance for cavitation if the entrance roundness of orifice is less.
- For $r/d = 0.16$, there is no vapour formation or cavitation when the cavitation number is greater than or equal to 1.41.
- For $r/d = 0.12$ and 0.16, similar results are obtained, so further increase in entrance roundness may not bring any improvement.
- In order to have minimum cavitation, the entrance roundness of $r/d = 0.12$ and above can be recommended.

Acknowledgements We are grateful to Dr. N. Unnikrishnan, Prof. Reeba Thomas and Prof. Smitha Mohan K., Department of Civil Engineering Trichur, for their generous support, patience, guidance, knowledge, time and enthusiasm and expert advice for the execution and completion of the work.

References

1. Chen, Y., & Heister, S. D. (1995). Two-phase modelling of cavitated flows. *Computers & Fluids, 24*(7), 799–809.
2. Darbandi, M., Sadegh, H., & Schneider, G. E. (2009). Simulating orifice cavitation using the full cavitation model. In *17th Annual Conference of the CFD Society of Canada*.
3. Nurick, W. H. (1976). Orifice cavitation and its effect on spray mixing. *Journal of Fluids Engineering*.
4. Yang, P. (2015). *Numerical study of cavitation within orifice flow*. Graduate and Professional Studies of Texas A&M University.
5. Tukiman, M. M., Ghazali, M. N. M., Sadikin, A., Nasir, N. F., Nordin, N., Sapit, A., & Razali, M. A. (2017). CFD simulation of flow through an orifice plate. *IOP Conference Series: Materials Science and Engineering, 243*.
6. Darbandi, M., & Sadeghi, H. (2010). Numerical simulation of orifice cavitating flows using two-fluid and three-fluid cavitation model. In *International Journal of Computation and Methodology*.

Flood Risk Assessment Methods—A Review

Ginu S. Malakeel, K. U. Abdu Rahiman, and Subha Vishnudas

Abstract Floods are the most prevalent natural disasters worldwide. Studies have shown that the frequency of flooding will increase in the future. Flood mitigation measures, which include structural as well as non-structural measures, can be selected effectively once the risk assessment for the area has been done accurately. In this paper, a review on flood risk assessment methods has been done considering flood risk as a function of flood hazard and vulnerability. This paper intends to analyse some of the most popular methods, including statistical methods, GIS and remote sensing, for assessing the flood hazard, flood vulnerability and flood risk, along with relevant case studies.

Keywords Flood · Flood risk · Hazard · Vulnerability · Hydraulic modelling · GIS · Satellite remote sensing · Climate change

1 Introduction

In recent years, humans have encountered increasing number of natural disasters, of which flood is the greatest and most common throughout the world [1]. Floods can be explained as excess flows exceeding the transporting capacity of river channel, lakes, ponds, reservoirs, drainage system, dam and any other water bodies, whereby water inundates outside water bodies areas [2].

G. S. Malakeel (✉)
Civil Engineering Department, Cochin University of Science and Technology, Kochi, India
e-mail: ginusaji@gmail.com

K. U. Abdu Rahiman · S. Vishnudas
Faculty, Civil Engineering Department, Cochin University of Science and Technology, Kochi, India
e-mail: arku@cusat.ac.in

S. Vishnudas
e-mail: v.subha@cusat.ac.in

Flooding is caused by several factors and is always preceded by heavy intense rainfall. Generally, floods can be classified into four: coastal, river, flash and urban floods [3]. Coastal flooding is due to the rise in sea levels resulting from extreme weather and high tides. River flooding is one of the most common types of inland flood, occurring along the banks of a river, because of heavy rainfall over a prolonged period of time. When the intensity of rainfall exceeds the infiltration rate, the ground cannot absorb the water as quickly as it falls and results in flash flooding. Extreme local rainfall coupled with blockage of urban drainage system is responsible for urban flooding.

Floods are likely to happen more rigorously and regularly down the line because of climate change, unplanned rapid urbanization, improper watershed management and reduction in the infiltration into groundwater by extension of impermeable surfaces in urban areas [4]. Thus, even if natural factors like climate change can be blamed for the increased probability of occurrence of floods, we cannot set aside the anthropogenic factors which are the key attributes. An increase in the frequency of floods resulted in severe losses which include lives of people, damage to property and infrastructure, as well as the destruction of the natural environment.

Recent major flood events such as those experienced in Bombay, Chennai and Kerala in 2018 and 2019 highlighted the serious threats posed by flooding. Hence, to make strategies for preventing and reducing flood damage, the study on risk assessment and zoning of flood prone areas caused by heavy rainfall is significant.

Flood risk is defined as "the probability of harmful consequences, or expected losses (deaths, injuries, property, livelihoods, economic activity disrupted or environment damaged) resulting from interactions between natural or human-induced hazards and vulnerable conditions" [5]. Mathematically, it is the product of flood hazard and flood vulnerability. Flood hazard is the physical and statistical aspects of the actual flooding (e.g. return period of the flood, extent and depth of inundation and flow velocity); whereas, flood vulnerability is the exposure of people and assets to floods and the susceptibility of the elements at risk to suffer from flood damage [6]. Even though there are a wide varieties of flood risk assessment methods, this paper intends to have an insight into the flood risk assessment methods considering flood risk as the product of flood hazard and flood vulnerability.

2 Flood Hazard Assessment

The United Nations Department for Humanitarian Affairs (DHA) defines hazard as "a threatening event, or the probability of occurrence of a potential damaging phenomenon within a given time period and area". The term "potentially damaging" means the phenomenon necessarily need not cause damages to the exposed elements, but it can [7]. Flood hazard is the probability that a flood event of a certain magnitude will occur in a given period of time in a given area [8]. The aim of flood hazard assessment is to determine the probability of flood exceeding a particular intensity over an extended period of time. Flood hazard assessment can be considered to be an amalgamation of two steps: determination of flood characteristics and determination of the probability of occurrence of each flood. The flood characteristics include extent, depth and flow velocity; whereas, the probability determination aims to link the flood characteristics to a recurrence interval.

2.1 Estimation of Design Flood

Initial step in the determination of flood characteristics is the estimation of design flood. There are three basic approaches to the estimation of design flood: [9].

(a) Hydro-meteorological approach
(b) Statistical approach
(c) Use of empirical formulae.

Design flood estimation by hydro-meteorological approach mainly involves the estimation of probable maximum flood by applying unit hydrograph theory. This method usually involves the derivation of a unit hydrograph from the assessment of design storm hyetographs. The main advantage of this approach is that it gives complete hydrograph and it allows making realistic estimation of its moderating effect while progressing through a river reach. Sahu et al. [10] derived an equation for design flood estimation in Baitarani basin (Odisha) on the basis of design flood estimated by hydro-meteorological method after considering catchments of ten reservoir projects of different sub-basins of Baitarani basin. Further, Sahu et al. [11] have worked on estimation of design flood for water resources projects for ungauged catchments of Rushikulya basin on design flood estimated by hydro-meteorological method considering eight existing/proposed reservoir projects on different tributaries in the basin. This approach, however, has certain limitations such as (a) flow records are seldom available at structure sites (b) self-recording rain gauge data does not exist for most of the water sheds for the concurrent period of the flow under consideration (c) assumptions in the derivation of the unit hydrograph for the catchment may not be satisfied always and (d) long-term hydro-meteorological data for estimation of design storm parameters are mostly not available [10, 11].

Statistical approach, commonly known as flood frequency analysis approach, relies on the existence of long flood records. Usually, the highest recorded flood from each year is used in the analysis. These data points are referred to as annual discharge maxima. Once these data points have been identified, a convenient probability distribution (e.g. log-normal, log-Pearson or generalized extreme value) is applied to fit the data set [12]. The selection of an appropriate distribution will be difficult with regions having small flood records. Azhar Husain [9] has estimated the design flood discharge for different return periods at Kakkadavu dam site in Kariangode river basin using Gumbel's and log-Pearson methods. He found that for return periods upto 5 years Gumbel's method provide higher values of peak discharge; whereas, for high return periods, log-Pearson method produces higher values [9]. It must be emphasized that proper application and interpretation of statistical procedures require substantial experience and specialized knowledge. Also, the accuracy of this method depends on the volume of data available [13].

Frequency analysis and hydro-meteorological methods of design flood estimation require adequate field observations and data collections. In the absence of such data, use of empirical formulae serves the purpose. General form of an empirical equation is given by $Q = CA^n$, where Q is the peak flow, C is a constant for the catchment, A is the catchment area, and n is an index. Therefore, the major parameter which is considered in this method is the catchment area and all the other parameters that influence the peak flow, like catchment and storm characteristics, are merged in the constant C. Empirical formulae such as Dicken's formula, Ryves formula and Inglis formula are applicable, respectively, to Northern India, Western India and Southern India. Jena and Nath [14] developed an empirical formula for the computation of design flood of ungauged catchments in Brahmani Basin of Odisha by establishing a relation between the peak discharges of the unit hydrograph and the design discharges of eight numbers individual sub-basins. The major limitation of applying the empirical formula is that they cannot be applied universally. Fixing the value of C is another major constraint in this approach of determining the peak discharge [14].

2.2 Hydraulic Modelling

Once the design flood has estimated, the next step in the flood hazard assessment is the estimation of the flood characteristics. Estimation of the flow extend, depth and flow velocity from the estimated design flood is done through hydraulic modelling or using GIS, which in turn provides flood hazard maps. There are a large number of hydraulic models, mostly based on solving hydrodynamic equations, in practice, and the selection of the model depends on the available data, location of the study area and the level of accuracy required. The results obtained from hydraulic models have to be calibrated and validated, using the flood data from real-life scenarios. Generally, the models can be categorized into three: one-dimensional (1-D) models, two-dimensional (2-D) models and three-dimensional (3-D) models.

One-dimensional models—They solve full or part one-dimensional St.Venant's continuity and momentum depth-averaged equations in the longitudinal direction [15]. One-dimensional hydraulic model defines the channel cross sections and solely considers the flow velocities perpendicular to the defined cross sections. They do not require the full stream bathymetry but only stream cross sections to describe the stream geometry. Even if they cannot represent complex flow patterns, they are still very useful as they are computationally efficient, and they can simulate first-order conditions over much larger stream domains and over much longer periods than 2D models [16]. Commonly used 1-D model is HEC-RAS, developed by the U.S. Army Corps of Engineers. It can employ 1-D flood routing in both steady and unsteady flow conditions with low computational cost, and it is public domain software [15]. Khattak et al. [17] assess the suitability of HEC-RAS model, in combination with

GIS, to determine the extent of inundation under different return period floods for Kabul River in Pakistan. Comparison of simulation of 2010 flood with the image of the flood taken by MODIS clearly shows a close agreement between the two [17]. GIS-based visualization of the flood hazard areas is a cost-effective approach to provide warning for people in flood prone regions [18].

Two-dimensional models—They solve full or part one-dimensional St.Venant's continuity and momentum depth-averaged equations in different directions rather than along the perpendicular direction to the cross sections [17]. That is, they use the grid or mesh-based depth-averaged approach for modelling. They can provide a map of flow properties such as water surface elevation, water depth, depth-averaged velocity and bottom shear stress [12]. They have been applied mostly for steady-state conditions on short reaches, since they are computationally expensive [14, 19]. Ghimire [20] assessed the flood risk from extreme storm events in the Tarland Burn in Aberdeenshire, North East Scotland, by the application of TUFLOW, a 2-D hydrodynamic model. The model has used digital terrain model obtained from fine-scale LIDAR to represent the flood plain topography. The model was tested using a historical flood event which had a return period of 1 in 2 years. It has been found that there was a good match between the modelled and the observed flood extents [20].

Three-dimensional models provide a scope of understanding the flow interaction with hydraulic structures which is commonly present in a river [21]. They use structured or unstructured grid or mesh-based approach to solve complex riverine systems where flow around structures, depth varied velocities, etc., exist. Delft3D is a fully integrated 3D modelling computer software technique applied to simulate hydrodynamics, sediment transport, waves, morphological developments, water quality and ecology for fluvial, estuarine and coastal environments [22]. Kumbier et al. [23] investigated compound flooding in an estuary, in Australia, using Delft3D model. They estimated coastal flood risk resulting from a separation of storm tide and riverine flooding processes and by comparison with the simulation results including and excluding riverine discharge demonstrated large differences in modelled flood extents and inundation depths. They concluded that coastal flood risk should be assessed by considering storm tide and riverine flooding drivers jointly [23]. Three-dimensional models require long run time and obviously they have more computational burden compared with 1-D and 2-D models.

The accuracy of the results obtained from the hydraulic models relies largely on the input data, the DEM in particular. Studies have shown that the quality and accuracy of the DEM are more relevant than the resolution and precision of the DEM, to support flood inundation models [24].

GIS and satellite-based remote sensing can be used to prepare flood hazard risk zone maps using diverse methods. In multi-criteria assessment method, thematic maps of rainfall distribution, drainage density, land use, soil type, size of micro watershed, slope, and roads per micro watershed can be prepared and by applying weights and ranks to these thematic maps, and the weightage maps can be prepared. The flood hazard risk zone maps prepared using weighted overlay analysis method will show the total areas subjected to the hazards, as very low, low, moderate, high and very high-risk zones. GIS offers a very decent display by combining, manipulating and analysing the information for the assessment of probable flood risk extents very swiftly and more proficiently [8]. The major difficulty of the weighted overlay analysis method lies in the determination of weights [25].

3 Flood Vulnerability Assessment

The other major component of flood risk assessment is flood vulnerability. FEMA defines vulnerability as the measure of the ability to weather, resist or recover from the impacts of a hazard in the long term as well as the short term. Accordingly, the hazard can be same for an area but the vulnerability depends on many factors such as land use, extent and type of construction, contents and use, the nature of populations (mobility, age, health) and warning of an impending hazardous event and willingness and ability to take responsive actions. Generally, flood vulnerability is described as the degree of damage that can be caused by a flood calamity. The definition of flood vulnerability is firmly rooted in how people or societies are likely to be affected by flood phenomena—that is, the sensitivity of the community or people to flooding considering the socio-economic, environmental and physical components [26]. The primary methods can be grouped into three categories those based on disaster loss data, the vulnerability index system and the vulnerability curve. Integrating with GIS, flood vulnerable maps can be prepared from the above-mentioned methods.

3.1 Flood Vulnerability Index (FVI) Method

This is the most commonly used method for flood vulnerability assessment. The vulnerability index method depends on complicated indices and their subjective weighting, to assess the vulnerability of the region to the floods. This method when coupled with GIS helps decision-makers to identify flood prone areas more easily and efficiently. The limitation of this method lies in the standardization of weights. Balica et al. [27] determined a flood vulnerability index (FVI), based on four components of flood vulnerability: social, economic, environmental and physical and their interactions. The conceptual FVI equation they have used is

$$FVI = Exposure \times Susceptibility/Resilience \qquad (1)$$

FVI of each of the above components are calculated separately, and the average of all the FVI components yields the total FVI. The study concluded that FVI is useful in a larger-scale vulnerability assessment when compared with deterministic approach [27]. Typical designations of flood vulnerable indices are shown in Table 1.

Table 1 Typical flood vulnerability designations

Designation	Index
Very small	<0.01
Small	0.01–0.25
Vulnerable	0.25–0.50
High vulnerability	0.50–0.75
Very high vulnerability	0.75–1

Balica et al. [27]

3.2 Vulnerability Curve Method

The vulnerability curve, also known as stage-damage curve or depth-damage curve, is developed based on the actual damage survey. Depth-damage relationship presents information on the relationship of flood damage of a certain element to a certain depth of flooding. Baky et al. [22] assessed the flood vulnerability for different land uses in the Baniachong Upazila (Sub-district), one of the flood-affected Upazilas in Bangladesh, using the vulnerability curve method. Here, initially the elements at risk are identified and then the depth-damage relationship of each of these elements is identified separately by collecting the flood damage data from secondary literatures and organizations and by conducting surveys which include extensive interviews with the local people as a part of the questionnaire survey. The information for all samples of each element class is then averaged and stage-damage curves are developed. The vulnerability curve prepared for the study area is shown in Fig. 1 [22]. The vulnerability curve method, therefore, depends on an extensive survey that takes a lot of time and resources. Additionally, the vulnerability curve for one region is not necessarily applicable to other regions.

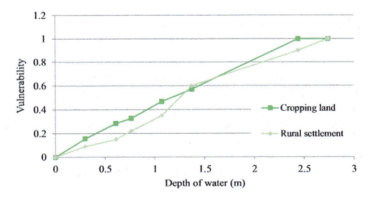

Fig. 1 Vulnerability curve of cropping land and rural settlement for Baniachong Upazila. *Source* Baky et al. [22]

3.3 Method Based on Disaster Loss Data

By and large, the flood losses can be divided into two: direct losses and indirect losses. Direct losses of flood are those effectuated by the direct contact of a flood event, whereas indirect losses are those which happen as the after effects of a flood event [28]. This method is based on the losses (both direct and indirect) that have occurred due to actual flood events. Huang et al. [29] had identified multidimensional flood vulnerability—population, death, agriculture and economy—using flood damage data and socio-economic statistical data from 2001 to 2010 at the provincial scale in China. The four models of flood vulnerability assessment were developed based on the data envelopment analysis method [29]. The method is simple; however, the assessment results should be treated with caution because disaster loss data is often unevenly recorded or inaccurate.

4 GIS Method

Flood vulnerability maps created using Geographic Information System (GIS) can be used to identify areas that are potentially prone to flash, urban and fluvial floods. GIS can be used in conjunction with a wide range of methods to obtain the flood vulnerability map. Feloni et al. [30] have done a flood vulnerability assessment of the Attica Region using a GIS-based multi-criteria approach. They have selected criteria appropriate for the study area, standardized and classified the selected criteria, and finally suitable weights were assigned to them. This method is extremely useful in areas where the other methods fail due to limited data availability, especially in ungauged catchments [30, 31].

5 Flood Risk Assessment

Flood risk is a function of hazard and vulnerability [30]. It is the probability of loss due to flood of a given intensity. Flood hazard is the same for a given area in terms of intensity but the risk could be different which depends on vulnerability [22]. Since flood risk is a combination of hazard and vulnerability, risk assessment can be achieved without much effort, once the hazard and vulnerability assessments have been completed. Mathematically,

$$\text{Flood Risk} = \text{Flood Hazard} \times \text{Flood Vulnerability}$$

Baky et al. [22] computed floodplain inundation depths for hazard assessment using Delft3D model. Elements at risk were identified, and flood damage functions were assessed, for the flood vulnerability assessment. In the final step of risk assessment, the expected damage of the risk element was estimated using the following equations.

$$D = \text{Vul} \times P \times A \tag{2}$$

$$ED = \text{Probability} \times D = (1/T) \times D \tag{3}$$

where D is total direct property damage per cell of the raster map, "vul" is the vulnerability value per cell which is the function of depth in meter and duration of inundated land in days, A is the area of each cell in sq.m, and P is the property value in monetary terms of each cell. Here, ED is the expected damage, and T is the return period of food.

With the integration of GIS tools, the estimated expected damages were classified into several classes and the final flood risk map was prepared [22]. Flood risk map prepared by M. A. A. Baky et al. for the Baniachong Upazila sub-district is in shown in Fig. 2.

Thus, the steps involved in the flood risk assessment can be summarized as follows: Initially, the design flood discharges for various return periods are estimated, and then this design discharges are converted into corresponding flood stages, usually by hydraulic modelling, represented by flood hazard maps. Then, under the vulnerability

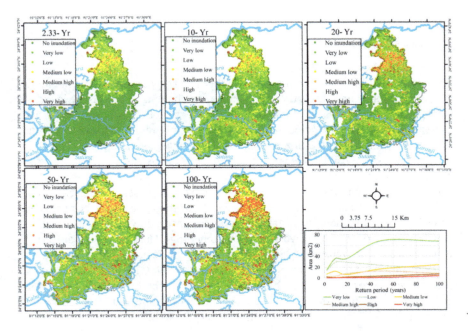

Fig. 2 Example for a flood risk map. *Source* Baky et al. [22]

assessment, the damage associated with each flood stage is estimated, and the final step is the flood risk assessment which encompasses the estimation of expected damage corresponding to various return periods. With the integration of GIS tools, the final flood risk map can be prepared.

Flood risk assessment from hazard and vulnerability assessments requires a tremendous amount of data which need to be accurate as well. Collection of such a large amount of data may become hardly possible in some scenarios, especially in urban areas. A time series analysis of satellite images for flood hazard and flood risk assessment was therefore developed and applied for the mapping of flood risk in urban areas. Here, the relative frequency of inundation (RFI) is mapped from the maximum flood extent images created from satellite images. The resulting RFI map can be used further for flood risk assessments. Even though the method has minimum data requirements, the quality of the RFI map relies on the satellite images. Further, this method cannot be useful in extrapolating water depths. On comparing, the flood hazard maps obtained from satellite images show 98% spatial agreement with that prepared from hydraulic modelling [32].

Event-specific flood hazard maps can be prepared from satellite imagery-derived flood extent maps using different image processing techniques. The flood risk map is further generated from the flood hazard map and vulnerability weighted land cover vector data in Geographic Information System environment. Although this method is appealing to flood relief planners and managements, it is worth noting that the method ignores two important parameters, flow velocities and flow depths [33].

6 Conclusions

This paper discusses flood hazard, vulnerability and risk assessment, including risk assessment approaches, methods, uncertainties and challenges. Floods are inevitable hydrologic phenomenon, and absolute flood risk protection is not possible. Risk assessment deals with the perception and evaluation of previous, current and future events of flood. Flood mitigation measures, which include structural as well as non-structural measures, can be selected effectively once the risk assessment for the area has been done accurately.

As we have seen, several approaches for flood hazard assessment exist. Each method has its own advantages and limitations. Even though most of the researchers prefer hydrologic models for hazard assessment, uncertainty is a major issue that needs to be encountered in every step of modelling. These models can provide accurate results only if the input data required is available in good quality and quantity.

In most of the studies, FVI method in association with GIS has found to provide an effective tool for flood vulnerability assessment. Vulnerability curve method and the method based on historical loss data require lot of time and effort, since their data requirement is more in comparison with FVI method.

One of the main constraints in flood risk assessment is the non-availability of reliable flood data. This difficulty can be alleviated, to some extent, by the use of satellite images and GIS. Again, flood risk as a function of flood hazard and vulnerability is only considered here, even though several other descriptions of risk assessment exist.

References

1. Dilley, et al. (2005). *Natural disaster hotspots: A global risk analysis. disaster risk management*, Series No. 5–34423. Washington DC: The World Bank.
2. Aris, M. M. (2003). *GIS modelling for river and tidal flood hazards in Waterfront City: Case Study in Samurangu City*, Java, Indonsia. ITC, Netherlands, Thesis Report
3. Rashetnia S. (2016). *Flood vulnerability assessment by applying a fuzzy logic method: A case study from Melbourne*. Thesis Report
4. Nasiri, H., Yusof, M. J. M., & Ali, T. A. M. (2016). An overview to flood vulnerability assessment methods. *Journal of Sustainable Water Resources Management, 3*.
5. ISO/IEC Guide 73:2002, Risk management—Vocabulary ñ guidelines for use in standards
6. Vojtek, M., Vojteková, J. (2016). Flood hazard and flood risk assessment at the local spatial scale: A case study geomatics. *Natural Hazards and Risk, 7*(6), 1973–1992
7. Gouldby, B., & Samuels, P. (2005). Integrated flood risk analysis and management methodologies, Floodsite, Language of Risk, Project Definitions, Report: T32-04-01
8. Ajin, R. S., Krishnamurthy, R. R., Jayaprakash, M., & Vinod. P. G. (2013). Flood hazard assessment of Vamanapuram River Basin, Kerala, India: An approach using remote sensing & GIS techniques. *Advances in Applied Science Research, 4*(3), 263–274
9. Husain, A. (2017). Estimation of design flood discharge for Kakkadavu Dam in Kariangode River Basin. *International Journal of Environmental Science & Natural Resources, 4*(1), 555626
10. Sahu, G. C., Prasad, C. V, Jena, J. G., Ray, G. P., & Das, A. K. (2005). Derivation of an equation for estimation of design flood for water resources project in Baitarani Basin (Orissa). In 46th Technical Annual, Institution of Engineers (India), Odisha Chapter, Bhubaneswar
11. Sahu, G. C., Prans, C. V., Jena, J. G., Das, A. K.(2006). Estimation of design flood of water resources projects of under gauged catchments in Rushikulya Basin (Orissa). In 47th Technical Annual, Institution of Engineers (India) (pp. 224–229)
12. *Methods in flood hazard and risk assessment*, Technical Notes, CAPRA
13. Karim, M. A., & Chowdhury, J. (1995). A comparison of four distributions used in flood frequency analysis in Bangladesh. *Hydrological Sciences Journal, 40*(1), 55–66. https://doi.org/10.1080/02626669509491390
14. Jena, J., & Nath, S. (2019). An empirical formula for design flood estimation of un-gauged catchments in Brahmani Basin, Odisha. *Journal of Institution of Engineers (India) Series A*. https://doi.org/10.1007/s40030-019-00402-x
15. Dimitriadis, P., Tegos, A., Oikonomou, A., Pagana, V., Koukouvinos, A., Mamassis, N., et al. (2016). Comparative evaluation of 1D and quasi-2D hydraulic models based on benchmark and real-world applications for uncertainty assessment in flood mapping. *Journal of Hydrology, 534*(2016), 478–492.
16. Benjankar, R., Tonina, D., & McKean, J. (2014). One-dimensional and two-dimensional hydrodynamic modeling derived flow properties: Impacts on aquatic habitat quality predictions. *Earth Surface Processes and Landforms*.
17. Khattak, M. S., Anwar, F., Saeed, T. U., et al. (2016). Floodplain mapping using HEC-RAS and ArcGIS: A case study of Kabul River. *Arab J Sci Eng, 41*, 1375. https://doi.org/10.1007/s13369-015-1915-3

18. Alho, P., & Aaltonen, J. (2008). Comparing a 1D hydraulic model with a 2D hydraulic model for the simulation of extreme glacial outburst floods. *Hydrological Processes, 22*(10), 1537–1547. https://doi.org/10.1002/hyp.6692
19. Leedal, D., Neal, J., Beven, K., Young, P., & Bates, P. (2010). Visualization approaches for communicating real-time flood forecasting level and inundation information. *Journal of Flood Risk Management, 3*, 140–150.
20. Ghimire, S. (2013). Application of a 2D hydrodynamic model for assessing flood risk from extreme storm events. *Climate, 1*, 148–162. https://doi.org/10.3390/cli1030148
21. Peruzzi, C., Castaldi, M., Francalanci, S., Solari, L. (2018). Three-dimensional hydraulic characterisation of the Arno River in Florence. *Journal of Flood Risk Management, 2018*, e12490. https://doi.org/10.1111/jfr3.12490
22. Baky, M. A. A., Islam, M., & Paul, S. (2019). Flood hazard, vulnerability and risk assessment for different land use classes using a flow model. *Earth Systems and Environment*. https://doi.org/10.1007/s41748-019-00141-w
23. Kumbier, K., Carvalho, R., Vafeidis, A., & Woodroffe, C. (2018). investigating compound flooding in an estuary using hydrodynamic modelling: A case study from the Shoalhaven River, Australia. *Natural Hazards and Earth Systems Sciences, 18*, 463–477.
24. Ali, A., Solomatine, D., Baldassarre, G. (2015). Assessing the impact of different sources of topographic data on 1-D hydraulic modelling of floods. *Hydrology and Earth System Sciences, 19*, 631–643.
25. Kourgialas, N., & Karatzas, G. (2011). Flood management and a GIS modelling method to assess flood-hazard areas—A case study. *Hydrological Sciences Journal, 56*(2).
26. Munyai, R. B., Musyoki, A., & Nethengwe, N. S. (2019). An assessment of flood vulnerability and adaptation: A case study of Hamutsha-Muungamunwe Village, Makhado Municipality. *Journal of Disaster Risk Studies, 11*(2), a692. https://doi.org/10.4102/jamba.v11i2.692
27. Balica, S. F., Popescu, I., Wright, N. G., et al. (2013). Parametric and physically based modelling techniques for flood risk and vulnerability assessment: A comparison. *Environmental Modelling and Software, 41*, 84–92. ISSN 1364-8152
28. Merz, B., Kreibich, H., Schwarze, R., & Thieken, A. (2010). Review article "Assessment of economic flood damage." *Natural Hazards and Earth Systems Sciences, 10*, 1697–1724.
29. Huang, D., Zhang, R., Huo, Z., et al. (2012). An assessment of multidimensional flood vulnerability at the provincial scale in China based on the DEA method. *Natural Hazards, 64*, 1575–1586. https://doi.org/10.1007/s11069-012-0323-1
30. Feloni, E., Mousadis, J., & Baltas, E. (2019). Flood vulnerability assessment using a GIS-based multi-criteria approach—The case of Attica region. *Journal of Flood Risk Management, 13*, (Suppl. 1), e12563. https://doi.org/10.1111/jfr3.1256
31. Ahmed, F. (2018). Flood vulnerability assessment using geospatial techniques: Chennai, India. *Indian Journal of Science and Technology, 11*(6). https://doi.org/10.17485/ijst/2018/v11i6/110831
32. Skakun, S., Kussul, K., Shelestov, A., & Kussu, O. (2014). Flood hazard and flood risk assessment using a time series of satellite images: A case study in Namibia. *Risk Analysis, 34*(8).
33. Schumann, G., & Di Baldassarre, G. (2010). The direct use of radar satellites for event-specific flood risk mapping. *Remote Sensing Letters, 1*(2), 75–84.

Flood Hazard Assessment and Flood Inundation Mapping—A Review

Reshma Antony, K. U. Abdu Rahiman, and Subha Vishnudas

Abstract Flood is one of the most common destructive natural disasters. The damage due to flood is huge and irreparable. Flood hazard is a combination of several factors which include both natural and manmade. Flood inundation map is one of the most essential tools which help urban and infrastructure planners for the future development of city. In this paper, a review is done on various methods of flood hazard assessment and flood inundation mapping using the latest technologies which integrate Geographical Information System (GIS) and various hydrodynamic models. The paper comprises of the case studies from different parts of the world.

Keywords Flood hazard · Flood inundation · GIS · Analytic hierarchy process · Weighted overlay

1 Introduction

Flood is defined as a condition when water temporarily rises to an unusual height above normal level for such water course, lake, or ocean [1]. Floods are usually expressed by physical characteristics such as flood depth, extend, and duration. Reducing vulnerability and flood risk is a key goal of flood risk management. Vulnerability is the degree to which system is susceptible to the adverse effects. But flood risk can be expressed as a number that is related to flood damage that happens to human, businesses, and nature [2]. Flood hazard may be defined as the exceedance

---

R. Antony (✉)
Faculty in Department of Civil Engineering, Sahrdaya College of Engineering and Technology, Kodakara, India
e-mail: reshmaantony@cusat.ac.in

K. U. A. Rahiman · S. Vishnudas
Faculty of Civil Engineering, School of Engineering, Cochin University of Science and Technology, Cochin, Kerala 682022, India
e-mail: arku@cusat.ac.in

S. Vishnudas
e-mail: subhakamal@gmail.com

probability of potentially damaging flood situations in a given area within a given period of time [3]. Flood hazards and disasters are the products of an interaction between environmental and social processes. Although these events are of natural origin, they are also caused by social, economic, and political environment which structures and configures the lives of individual and group of people [4]. It was estimated that a total of nearly 2.8 billion people were affected by floods during time frame of 1980–2009 [5].

The floodplain demarcation for different flood magnitudes is needed to lessen the damage to infrastructure and loss of lives [6]. High reservoir storage, which in turn results in the release of excess water and abnormal extreme precipitation in the catchments, upstream to the major reservoirs, might also contribute to large scale flooding [7]. The impacts of landuse and landcover (LULC) change and urbanization on flooding are also not minor. A study conducted in Oshiwara River Basin in Mumbai revealed that the spatial and temporal changes in the land use over the four decades resulted in a remarkable increase in the runoff peak and volume [8].

Flood risk map is a useful tool to the decision-makers for probable defensive measures, better landuse planning and flood risk management in this changing climate [9, 10]. Multi-sensor data can be very effectively used for flood monitoring purposes and flood hazard management so that the loss of life and related damages can be reduced to a large extends [11]. Stringent actions should be taken regarding the uncontrolled urbanization and the occupation of areas that are very close to rivers and to be executed by policymakers in order to put off more major damages [9]. Saaty in 1980 [12] introduced analytic hierarchy process (AHP) that can be very efficiently used for decision making in multi-criteria analysis problems. A better understanding about the contribution of each flood affecting factors is possible in AHP methodology, based on the weights given to them [9]. AHP allows a framework for group participation in problem-solving. It is a process of "systematic rationality," and it enables us to study the simultaneous interaction of various factors within a hierarchy [13]. Sahoo et al. [10] established the effectiveness of spatial technology and flood risk mapping in identifying the vulnerable areas of an urban catchment. Flood hazard maps help the authorities to figure out the inundation characteristics of floodplains. Islam and Sado [14] developed a land development priority map integrating flood hazard map and population density map for implementing flood mitigation measures and relief operations in high-risk areas. Most of all, high-resolution satellite imagery is an inevitable factor to assure the accuracy of flood hazard study [15].

2 Flood Hazard Map—Important Factors

The following factors are prerequisite for the generation of flood hazard map.

2.1 Landuse and Landcover Change

In flood estimation, especially in urban areas, the effect of landuse and landcover change and urbanization plays a significant role. Urbanization increases the imperviousness of ground surface, which in turn increases the surface runoff resulting in severe flooding during monsoon season [16]. Landuse and landcover change can cause an increase in peak discharge for higher return periods, and also, it increases the flood inundated area in the catchment [8, 16]. Experts have asserted that deforestation is the basic cause behind flood problems. A complete vegetation cover helps in reducing flood through the detention of rainfall by interception, increase infiltration, and reduce runoff by enhancing evaporation and evapotranspiration [17].

2.2 Rainfall Intensity

Precipitation is an important component of hydrologic cycle which bears a significant influence on hydrologic design and water resource management. Evaluation of extreme precipitation events in relation to intensity, duration, and influence of climate variability is significant to deal with the issues of floods and flooding mechanisms [18]. Extreme rainfall and climate variability have a potential role in almost all floods [6, 7, 19]. Climate change has great practical implications for river discharge variation and occurrence of extreme events of floods and droughts [20].

2.3 Drainage Density

Drainage density is defined as the cumulative length of all stream channels in a drainage basin, divide by the total drainage basin area. The hydrological response of a channel network is strongly influenced by drainage density [21]. Morphometric parameters, especially drainage density, are important in assessing the flash flood risk in an area [22]. A study conducted by Ameur [23] about floods in Jeddah, Saudi Arabia, indicates geomorphological specifics, rainfall and climate change as the main causes of floods. High drainage density, higher relief, and circulatory ratio have resulted in flooding of the Ogunpa Basin in Nigeria [24].

2.4 Slope and Elevation

Susceptibility to flooding decreases as the elevation increases [25]. Slope of the terrain is an important geomorphological parameter in assessing flood hazard. Topography of catchment especially in low land catchments is very important, and the flood parameters are highly susceptible to topographical changes [26]. A study conducted by Xie et al. [27] states that, the lower the terrain absolute height and the less the degree of terrain changes, the more it is vulnerable to flood disaster. Lower the slope gradient, the higher is the possibility of flooding and flood events [28].

2.5 Distance to the Main Channel

Regions located close to the main channel will have chance of getting flooded soon. This factor helps for future land development and for applying flood mitigation measures in areas at high risk [14].

2.6 Soil Permeability or Infiltration Characteristics

The higher the permeability of a soil layer, the more rapidly water can infiltrate through and keep away from flooding. A study conducted by Sampson [29] shows a direct correlation between soil permeability and flooded zone.

3 Case Studies

This section explains some case studies from elsewhere in the world. The section is divided into two subsections. First section discusses the methodologies adopted by researchers for generating flood hazard map, and second section mainly deals with case studies related to flood inundation mapping.

3.1 Flood Hazard Mapping

Ismail Elkhrachy conducted a study [30] to locate flood vulnerable areas in Najran City using Geographical Information System (GIS). In this study, the following factors were considered for flood hazard study: runoff, soil influences (soil type and texture), surface slope, surface roughness, drainage density, distance to main channel and land cover. The author used SPOT 5 DEM and SRTM DEM data for the study purpose, for which the quality was assured with ground control points using elevation obtained by GPS. River network, drainage basins, and landuse maps are extracted using ArcMAP toolbox. The soil conservation service curve number (SCS-CN) loss method is used to quantify runoff for the study area. Hydrological modelling is done by HEC-Geo HMS software. Surface roughness map is derived from landuse map. Analytic hierarchical process (AHP) was adopted for decision making. Pairwise comparison method is used to determine the relative importance of each factor. Each factor is divided into various classes, and weights are assigned from a minimum value of 1 to maximum value of 7. Flood Hazard Index was computed using weighted overlay analysis and raster calculator in ArcMap 10.1 software. The study is deficient in stream gauging records (only one station was available) and high-resolution DEMs for obtaining accurate results in this study.

Danumah et al. [9] developed a flood risk map of Abidjan District, by the integration of several elements under two criteria such as hazard and vulnerability. The study developed AHP flood hazard map by considering four factors: drainage density, soil type, slope, and isohyets. Also, AHP flood vulnerability map was prepared by taking in to account three elements such as urban structure types, population density, and drainage system. Pairwise comparison is performed by scaling proposed by Saaty [12]. Hazard and vulnerability maps are prepared separately by overlay technique, and later on, flood risk map is developed by crossing them. ArcGIS raster calculator and spatial analyst tools were used for performing these processes. He argued that geospatial techniques are highly reliable in natural disaster assessment provided the high spatial resolution satellite data is available. The author is of opinion that standardization if approach like linear instead of natural jenks can be improved for map comparison and accuracy assessment process. The results can be improved by developing urban structure types using oriented-based image analysis (OBIA) using high-resolution images like Quick Bird, Ikonos, and Rapid Eye.

Ntajal et al. [31] did an assessment and mapping of social flood risk in the Lower Mono River Basin, West Africa. Here flood risk assessment comprises of various factors like hazard, exposure, vulnerability, and capacity. This study adopted a framework of [32, 33]. In this study, the flood hazard assessment was done based on hydrological analysis from SRTM (30 m) DEM and landuse classification of the area. Weighted overlay analysis in ArcGIS was used to integrate land use/land cover, slope, elevation, rainfall data, flow accumulation, and soil. The weightages for each factors were assigned based on discussions with experts. Rank sum overlay in ArcGIS (10.1) platform was used in this study to generate the flood risk map. Hazard, exposure, vulnerability, and capacity measures were given a percentage of influence based on expert knowledge. The study also quoted that, the likelihood of a flood increases as the elevation of a location decreases. The study also found that, the built-up areas tend to increase surface runoff when the given slope is steep and thus reduces the rate of infiltration of surface water, which in turn increases the physical vulnerability of people and forest ecosystems to flood risk. Here capacity assessment is done based on the availability of flood disaster training programs, early warning systems, availability of evacuation facilities, their awareness of flood disaster, etc. The study concluded that the mitigation of flood risk can be accomplished through managing the hazard, reducing the exposure and vulnerability and capacities of communities. The author argued that, developing flood mitigation strategies need a collective effort from the communities, institutions, organizations, and governments. The main limitation in this study is that the resilience of communities to flood risk is not covered in this study.

Ozkan et al. [15] developed a flood hazard map of Izmir Province, by considering several meteorological, hydrological, topographical factors such as flow accumulation, elevation, slope, land use and rainfall intensity in different time periods. In this study, the author used SRTM DEM which has 90 m of grid size resolution. The study integrated these entire factors by means of weighted overlay method in GIS to determine potential flood zones. Here, first of all, the mutual interaction ratios of the above five factors which cause flood are calculated. Here 1 point is assigned for main effects and 0.5 point assigned for minor effects, and factor weights (or rates) are quantified. Thematic maps are prepared in GIS platform and reclassified in to various risk levels like very high, high, medium, low, and very low by assigning weights ranging between 1 and 10. Finally, weighted rates and total weights for each factor are determined, and the percentage of influence of each is obtained. Based on these values, weighted factors maps and final flood hazard map are generated. The use of high-resolution DEM could increase the accuracy of the work.

Matheswaran et al. [34] conducted a study to identify the potential flood risk zones in Bihar for steering the flood insurance scheme in Bihar. The study used two methods for regionalization of flood risk area (i) cluster and (ii) hotspot analysis methods. In clustering method, the optimum number of clusters is selected based on the value of the silhouette coefficient. Hotspot analysis is a two-step process, which involves weighted overlay which is available in ArcMap software. Final flood risk map was validated by comparing with official estimates recorded by the government. Flood-affected population was also quantified by overlaying gridded population data on the flood maps.

Ajin et al. [35] created a flood hazard risk map using ArcGIS and ERDAS Imagine software. The study considered seven major factors such as drainage density, land use and land cover, soil type, size of micro-watershed, rainfall distribution, slope and roads per micro-watershed for his study. These factors were given appropriate weightage, and weighted overlay analysis in GIS was used to delineate the flood risk zone of Vamanapuram River basin. The study concluded that the scale and frequency of flood are likely to increase due to the change in climate which in turn brings high intensity of rainfall.

Kazakis et al. [36] developed a methodology named "FIGUSED" for identifying the flood-prone areas in Rhodope–Evros region, Greece. It is an index-based method which quantifies flood hazard index by considering seven major factors which influences flood exposure such as flow accumulation, rainfall intensity, geology, land use, slope, elevation, and distance from drainage network. This study used AHP for assigning weights for each criterion. A validation of this study was also done through a single parameter sensitivity analysis. After sensitivity analysis they concluded that elevation, distance from drainage network, and slope are having bigger influence in the study area. The study also mentioned that geology is the least affecting parameter.

3.2 Flood Inundation Mapping

Mokhtar et al. [37] assessed flood risk in the Ghamsar Watershed using DEM and Hec-Geo RAS. This study quantified the flood damage to human, infrastructure, and agriculture using Hydrologic Engineering Center's Flood Impact Analysis (Hec-FIA software). In this study, Log Pearson type 3 method was used for flood frequency analysis as this method can be used for extreme values and probability-weighted moment's method for calculating the distribution quintiles. Channel cross-sectional details are extracted from maps by importing the same in to ArcGIS software. Later on these details are imported into Hec RAS using Hec-Geo RAS extension. The geometric information such as stream alignment and digital elevation model, information on buildings and agricultural land use, a raster file of inundation, impact area was imported in to HEC-FIA software, for assessing flood damage.

Gain and Hoque [38] assessed flood risk of the eastern part of Dhaka City by considering flood hazard, vulnerability, and damage due to flood. Here damage estimation is restricted to direct monetary loss of available landuse classes. In this study, five extreme value distributions were compared based on goodness-of-fit tests. Flood frequency analysis was performed using Gumbel's distribution. Flood inundation was done in HEC–RAS and geographic information system (GIS) extension HEC-Geo RAS. The study developed vulnerability maps for different designed floods, by taking vulnerability as a function of inundation depth and inundation duration. Finally, total damage is quantified as the product of three factors such as vulnerability, property value in monetary terms per square meter and area of each cell size.

Sahoo et al. [10] developed a flood inundation map and measured the flood risk of a catchment in northeast India using spatial technology. Flood depth and inundated area, land use, and population density and road networks were the major factors considered for flood hazard study. The study highlighted that flood depth and inundation area is more important and sensitive factors when compared to other factors. In this study, runoff was simulated by using storm water management model (SWMM) developed by Environmental Protection Agency. The study established that high urbanization makes the area more vulnerable to flood. A three-dimensional matrix multiplication approach was used to arrive at the combined flood hazard rank (the method was taken from Islam and Sado 2002).

Joy et al. [39] mapped the flood inundated area of Meloor Panchayat for 2018 Kerala flood, using GIS and participatory GIS technologies. A questionnaire survey was conducted in the study area using participatory GIS approach and found that 53% of study area was inundated during that flood period. The study used participatory GIS method due to the unavailability of satellite imagery during the peak flood hour.

ArcGIS and remote sensing can be integrated with any newly developed hydrological and hydraulic models to give very good or accurate results and can be used to generate maps to set off any sufficient flood mitigation measures or appropriately designed flood warning and relocation systems under a broad disaster management control plan [16].

4 Conclusions

This paper summarizes review carried out in the area of recent techniques for mapping flood hazard and flood inundation areas. Seven major parameters are making a good contribution toward flood hazard. Factors such as landuse and landcover change, rainfall intensity, drainage density, slope and elevation, population density, distance to the main channel, and soil infiltration characteristics are highly influencing flood hazard. The study helped in understanding the significance of Geographical Information System (GIS) technologies and analytic hierarchy process (AHP), in flood management process. This review also explains how GIS can be coupled with some of the mostly used hydraulic and hydrologic models such as HEC RAS, HEC-Geo RAS, and HEC-FIA for flood inundation mapping.

References

1. Regulation of Flood Hazard Areas to Reduce Flood Losses, Volume 1, Parts I-IV, Water Resources Council U.S.
2. Zanuttigh, B., et al. (2015). *Developing a holistic approach to Assessing and Managing Coastal Flood Risk, Coastal Risk Management in a Changing Climate*, Butterworth-Heinemann Publications.
3. Begum, S., et al. (2007). Flood risk mapping at local scale: Concepts and challenges. In *Flood Risk Management in Europe: Innovation in Policy and Practice, Chapter 13* (pp. 231–252) Springer.
4. Parker, D. J. (2000). *Introduction to floods and flood management, floods* (Vol. 1, pp :3 –39). London and NewYork: Routledge, Taylor and Francis Group.
5. Doocy, S., et al. (2013). The human impact of floods: A historical review of events 1980- 2009 and systematic literature review. *PLOS Currents Disasters*.
6. Mishra, V., et al. (2018a). Hydroclimatological perspective of the Kerala flood of 2018. *Journal of the Geological Society of India, 92*, 645–650.
7. Mishra, V., et al. (2018b). The Kerala flood of 2018: Combined impact of extreme rainfall and reservoir storage, *Hydrology and Earth System Sciences Discussions*.
8. Zope, P. E., et al. (2016). Impacts of land use–land cover change and urbanization on flooding: A case study of Oshiwara River Basin in Mumbai, India. *CATENA, 145*, 142–154.
9. Danumah, J. H., et al. (2016). Flood risk assessment and mapping in Abidjan district using multi-criteria analysis (AHP) model and geoinformation techniques, *Geoenvironmental Disasters, 3*(10).
10. Sahoo, S.N., et al. (2017). Development of flood inundation maps and quantification of flood risk in an urban catchment of Brahmaputra river. *ASCE-ASME Journal of Risk and Uncertainty in Engineering Systems, Part A: Civil Engineering, 3*(1).
11. Prasad, A. K., et al. (2006). Potentiality of multi-sensor satellite data in mapping flood hazard. *Journal of the Indian Society of Remote Sensing, 34*(3).
12. Saaty, T. L. (1980). *The Analytic Hierarchy Process*. New York: McGraw Hill.
13. Saaty, T. L. (2012). *The analytic hierarchy process for decisions in a complex world* (3rd end.) RWS Publications.
14. Islam M. M., & Sado, K. (2002). Development priority map for flood countermeasures by remote sensing data with geographic information system. *Journal of Hydrologic Engineering, 7*(5), 346–355 (@ASCE)

15. Ozkan, S. P., et al. (2016). Detection of flood hazard in urban areas using GIS : Izmir Case. In *9th International Conference Interdisciplinarity in Engineering, INTER- ENG 2015, 8–9 October 2015* (Vol . 22, pp. 373–381). Tirgu- Mures, Romania, Procedia Technology.
16. Zope, P. E. (2015). Impacts of urbanization on flooding of a coastal urban catchment: A case study of Mumbai City, India. *Natural Hazards, 75,* 887–908.
17. Masoudian, M. (2009). *The topographical Impact on effectiveness of flood protection measures.* Kassel University Press.
18. Teegavarapu, R. S. V. (2012). *Floods in a changing climate: Extreme precipitation.* New York: Cambridge University Press.
19. Olanrewaju, R., et al. (2017). Analysis of rainfall pattern and flood incidences in Warri Metropolis, Nigeris. *Geography, Environment, Sustainability, 10*(4), 83–97.
20. Iqbal, M. S., et al. (2018). Impact of climate change on flood frequency and intensity in the Kabul River Basin. *Geosciences, 8,* 114.
21. Goudie, A. S. (2004). *Encyclopedia of geomorphology* (Vol. 1 & 2, pp. 277) Routledge: Taylor and Francis Group.
22. Syed, N. H., et al. (2017) ISPRS annals of the photogrammetry, remote sensing and spatial information sciences. In *4th International GeoAdvances Workshop, 14– 15 October 2017* (Volume IV-4/W4) Safranbolu, Karabuk, Turkey.
23. Ameur, F. (2016). Floods in Jeddah, Saudi Arabia: Unusual phenomenon and huge losses. What prognoses. In *FLOOD risk 2016—3rd European Conference on Flood Risk Management.*
24. Ajibade, L. T., et al. (2010). *Morphometric analysis of Ogunpa and Ogbere drainage basins.* Ibadan, Nigeria.
25. Niyongabire, E., et al. (2016). Use of digital elevation model. In A. Gis (Ed.) *For flood susceptibility mapping: Case of Bujumbura City, Proceedings, 6 th International Conference on Cartography and GIS, 13–17 June 2016,* Albena, Bulgaria. ISSN: 1314-0604.
26. Masoudian, M. (2011). Influence of land surface topography on flood hydrograph. *Journal of American Science, 7*(11).
27. Xie, L., et al. (2013). Correlation between flood disaster and topography: A case study of Zhaoqing City. *Journal of Natural Disasters, 22*(6), 240–245.
28. Rahmati, O., et al. (2016). Flood hazard Zoning In Yasooj Region, Iran, using GIS and multicriteria decision analysis. *Natural Hazards and Risk, Geomatics, 7*(3), 1000–1017.
29. Sampson, S. E. (2016) *The correlation between soil permeability and flooding in the Northeast sector of the Dog River Watershed.*
30. Elkhrachy, I. (2015). Flash flood hazard mapping using satellite images and GIS tools: A case study of Najran City, Kingdom of Saudi Arabia (KSA). *The Egyptian Journal of Remote Sensing and Space Sciences, 18,* 261–278.
31. Ntajal, J. (2017). Flood disaster risk mapping in the Lower Mono River Basin in Togo, West Africa. *International Journal of Disaster Risk Reduction, 23,* 93–103.
32. Davidson, R. (1997). *An urban earthquake disaster risk index, Report No. 121*, The John A. Blume Earthquake Engineering Center, Stanford.
33. Bolin, C., Cardenas, C., Hahn, H., & Vatsa, K. S. (2003). Natural disasters network: Comprehensive risk management by communities and local governments. Inter-American Development Bank.
34. Matheswaran, K., et al. (2019). Flood risk assessment in South Asia to prioritize flood index insurance applications in Bihar, India. *Geomatics, Natural Hazards and Risk, 10*(1), 26–48.
35. Ajin, R. S., et al. (2013). Flood hazard assessment of Vamanapuram River Basin, Kerala, India: An approach using remote sensing & GIS techniques. *Advances in Applied Science Research, 4*(3), 263–274.
36. Kazakis, N., et al. (2015). Assessment of flood hazard areas at a regional scale using an index-based approach and analytical hierarchy process: Application in Rhodope-Evros Region, Greece. *Science of the Total Environment, 538,* 555–563.

37. Mokhtari, F. et al. (2017). Assessment of flood damage on humans, infrastructure, and agriculture in the Ghamsar Watershed using HEC-FIA software. *Natural Hazards Review, 18*(3).
38. Gain, A. K., & Hoque, M. M. (2013). Flood risk assessment and its application in the eastern part of Dhaka City, Bangladesh. *Journal of Flood Risk Management*, 219–228.
39. Joy, et al. (2019). Kerala flood 2018: Flood mapping by participatory GIS approach. *Meloor Panchayat, International Journal on Emerging Technologies, 10*(1), 197–205.

CPSIA information can be obtained
at www.ICGtesting.com
Printed in the USA
BVHW012354081220
595246BV00001B/10